Stephen F. McCormick
University of Colorado, Denver

Multilevel Projection Methods for Partial Differential Equations

SOCIETY FOR INDUSTRIAL AND APPLIED MATHEMATICS

PHILADELPHIA, PENNSYLVANIA   1992

Printed by Capital City Press, Montpelier, Vermont.

Library of Congress Cataloging-in-Publication Data

McCormick, S. F. (Stephen Fahrney), 1944-
    Multilevel projection methods for partial differential equations / Stephen F. McCormick
        p. cm. — (CBMS-NSF regional conference series in applied mathematics ; 62)
    Includes bibliographical references and index.
    ISBN 0-89871-292-0
    1. Differential equations, Partial—Numerical solutions.
    2. Multigrid methods (Numerical analysis) I. Title. II. Series.
    QA377.M32 1992
    515'.353—dc20   91-39536

Contents

Preface

This monograph supplements my recent book, published in SIAM's Frontiers series (*Multilevel Adaptive Methods for Partial Differential Equations*, SIAM, Philadelphia, 1989). The basic emphasis of that book is on problems in computational fluids, so it focuses on conservation principles and finite volume methods. This monograph concentrates instead on problems that are treated naturally by "projections," for which I have in mind finite element discretization methods. The eigenvalue problem for a self-adjoint diffusion operator serves as a prime example: it can be discretized by posing the problem as one of optimizing the Rayleigh quotient over the unknown projected onto a finite element space. This monograph shows that the projection discretization methods lead naturally to a full characterization of the basic multilevel relaxation and coarsening processes. That is, if we use the projection concept as a guide, the only basic algorithm choices we will have to make in implementing multigrid are certain subspaces that will be used in relaxation and coarsening. All other major components (e.g., interlevel transfers, coarse-level problems, and scaling) are determined by the projection principle.

The Frontiers book and this monograph served as the basis for the lectures I gave at the CBMS-NSF Regional Conference on "Multigrid and Multilevel Adaptive Methods for Partial Differential Equations," which was held at George Washington University, Washington, DC, on June 24–28, 1991. The purpose of the conference was to introduce the participants to the multigrid discipline, primarily in terms of multilevel adaptive methods, and to stress general principles, future perspectives, and open research and development problems. In tune with these objectives, this monograph lays the groundwork for development of multilevel projection methods and illustrates concepts by way of several rather undeveloped examples. This monograph contains no serious treatment of theoretical results because it would be inconsistent with our objectives—which is fortunate because almost no theory currently exists for multilevel projection methods.

An outline of this monograph is as follows:

Chapter 1 includes motivation, notation and conventions, prototype problems, and the abstract discretization method.

Chapter 2, which is the core of this book, describes the abstract multi-level projection method, its formulation for both global and local grid applications, and various practical issues.

Chapter 3 describes the unigrid algorithm, which is an especially efficient computational tool for designing multilevel projection algorithms.

Chapter 4 develops several prototype examples of practical uses of these methods.

Chapter 5 identifies a few open research problems relevant to multilevel projection methodology.

The development here assumes a basic understanding of multigrid methods and their major components, including discretization, relaxation, coarse-grid correction, and intergrid transfers. An appropriate foundation for this understanding is given in [Briggs 1987].

I am endebted to the National Science Foundation for support of the Regional Conference under grant DMS-9015152 to the Conference Board of the Mathematical Sciences and George Washington University. I am also thankful for the support of my research, on which this monograph is based, by the Air Force Office of Scientific Research under grant AFOSR-86-0126 and by the National Science Foundation under grant DMS-8704169. I am especially endebted to the Conference Director, Professor Murli M. Gupta, who convinced me that his idea of having this conference was an excellent one, and whose leadership made it such a success. My gratitude also goes to several colleagues whose advice and guidance were essential to my completion of this monograph: Achi Brandt, Gary Lewis, Tom Manteuffel, Klaus Ressel, Ulrich Rüde, John Ruge, and Gordon Wade. I am also grateful to the typist, Mary Nickerson, whose skills and editorial suggestions greatly improved the final product. Finally, I could not have finished this work without the patience and support of my loving wife Lynda, and for that I thank her.

Stephen F. McCormick
University of Colorado, Denver

To Mom

CHAPTER 1

Fundamentals

1.1. Introduction.

This monograph should be considered a supplement to the recent SIAM Frontiers book [McCormick 1989] on multilevel adaptive methods. That book contains a basic development of multilevel methods focused on adaptive solution of fluid flow equations. As such, it concentrates on finite volume discretization and multilevel solution methods that are compatible with physical conservation laws. Since many of the approaches of that book are relatively well developed and analyzed, it contains both numerical examples and underlying theory. Historical notes and references are also included.

The emphasis of this monograph, on the other hand, is on what we call *multilevel projection methods*. The classical prototype is the standard fully variational multigrid method applied to Poisson's equation in two dimensions, using point Gauss-Seidel relaxation, bilinear interpolation, full weighting, and nine-point stencils. (See [Stüben and Trottenberg 1982; §§1.3.4, 1.3.5, and 2.4.2], for example.) The key to understanding our basic idea here is that each stage of this prototype algorithm can be interpreted as a Rayleigh-Ritz method applied to minimizing the energy functional, where the optimization is taken as a correction over the continuous space *projected* onto certain subspaces of the fine-grid finite element space. In the stage where relaxation is performed at a given point, the relevant subspace is the one generated by the "hat" function corresponding to that point. In the coarsening stage, the relevant subspace is the coarse-grid finite element space.

In general, the Rayleigh-Ritz approach, and the related Galerkin and Petrov-Galerkin discretization methods, will act as guides to our development of corresponding relaxation and coarse-grid correction processes. These multilevel components will be developed in a way that is fully compatible with the projection discretization method that defines the fine-grid problem. One of the attributes of this approach is that the fundamental structure of all of the basic multilevel processes are in principle induced by the discretization. This can substantially simplify multilevel implementation, and it often provides the foundation for a very efficient algorithm.

The central idea behind multilevel projection methods is certainly not new.

1

Specific forms for linear equations have actually been known since the early stages of multigrid development (cf. [Southwell 1935]), and they have recently been used for eigenvalue problems [Mandel and McCormick 1989] and constrained optimization [Gelman and Mandel 1990]. Moreover, the basic idea is so natural and simple that it undoubtedly has had at least subconscious influence on many multigrid researchers. Yet, in almost all cases, specific forms of the projection concept were posed only in a Rayleigh-Ritz setting (however, see [McCormick 1982], which used a Galerkin formulation) on global grids (however, see [McCormick 1984] and [McCormick 1985], which consider the locally refined grid case). In any event, this monograph represents the first attempt to formalize and systematize projection methods as a general approach, incorporating Rayleigh-Ritz, Galerkin, and Petrov-Galerkin formulations as well as global and locally refined grids. Actually, even the material developed here can be considered only as a first step in this direction. We do not as yet have a full set of general principles for guiding the choice of specific procedures, or a founding theory. Nevertheless, numerical results for a few examples suggest that these projection methods have significant potential. In any case, their consistency with discretization methods is a seductive premise for further exploration.

As a supplement to the Frontiers book [McCormick 1989], this monograph will be fairly brief. Included are certain parts of that book needed here for emphasis, and the central development will for the most part be self-contained. In fact, to some readers (especially those most familiar with finite elements), it would be best to start with this monograph. However, the Frontiers book should be consulted for background, history, related reference material, a few numerical illustrations (based on multilevel schemes that are generally not of projection type), and theory (that actually does apply to projection methods, if only in the linear self-adjoint case). *Those who are unfamiliar with multigrid methods and their basic concepts should begin by reading more introductory material (cf. [Briggs 1987]).*

One of the simplicities of the following development is that it is set primarily in the framework of subspaces of the relevant continuum function spaces. Thus, the discretization and solution methods are developed mostly in terms of finite-dimensional subspaces of the function spaces on which the partial differential equation (PDE) is defined. This avoids some of the cumbersome notation (e.g., intergrid transfer operators) necessary when the nodal vector spaces are considered. In fact, the abstract methods apply even in cases where grids are not the basis for discretization (e.g., spectral methods). However, for practical relevance, we will occasionally use the nodal representation to interpret the concepts in terms of two prototype problems we shall introduce.

Another simplification is to restrict our attention to spatial problems. Time-dependent equations may be treated in a natural way by the projection methods developed, similar to the finite volume approach described in the Frontiers book. However, there is no doubt that much is to be gained by exploiting further the special temporal character of these problems, and it is perhaps better to postpone their treatment until this avenue has been more fully pursued.

Yet another simplification is to assume homogeneous Dirichlet boundary conditions for the PDE. (We will assume throughout this monograph that they are automatically imposed on all functions and nodal vectors. Along with this premise, we assume that all grids are "open" in the sense that they do not contain boundary points.) This is a standard simplification made in the finite element literature, and we find it particularly useful here since: it avoids additional terms in the weak forms; it avoids questions regarding approximation of boundary conditions on a given grid; and it allows us to treat the admissible PDE functions and finite element functions in terms of the *spaces* and *subspaces* in which they reside. For the reader who wishes to consider inhomogeneous but linear boundary conditions, it is enough to determine how they are to be imposed on the finest level: coarse-grid corrections should satisfy homogeneous conditions, even when the PDE is nonlinear. Since the fine-grid treatment of inhomogeneous boundary conditions is a classical question of discretization, we need not deal with it here.

These simplifications are likely to leave some readers with a false sense of understanding of certain practical implications of the principles and methods we develop. For example, it will seem natural and simple for variational problems to adopt the principle that the coarse-level correction should be defined so that it does its best to minimize the objective functional. However, it may be very much another matter to discern what interlevel transfer operators and coarse-level problems this abstract principle induces, let alone what happens at the software level. In fact, there are usually a variety of ways to realize such principles that are theoretically equivalent but have rather different practical consequences. Since the focus of this monograph is on basic principles and concepts, we will allude only to a very few of these issues. Therefore, the reader interested in practical aspects, especially those intent on using the ideas developed here, should begin by translating these principles into terms associated with the nodal representation of the discretization. Understanding details of the related multilevel adaptive techniques as they are developed in the Frontiers book [McCormick 1989] may be useful for this purpose.

1.2.   Notation and conventions.

The notation will conform as much as possible to that of the Frontiers book. However, the emphasis of that book is on finite volume methods for physical conservation laws, which means that a large percentage of the development involves finite-dimensional Euclidean spaces, and only a small percentage is devoted to continuum quantities. This is just the reverse of the present situation, which stresses finite element subspaces (i.e., continuum quantities) and only occasionally references nodal vector spaces. We therefore abandon the use of Greek symbols for the continuum and instead use underbar to distinguish Euclidean quantities.

Following is a general description of the notation and conventions, together with certain specific symbols, acronyms, and assumptions, used in this monograph.

Regions and grids use capital Greek letters; spaces, operators, and grid points use capital Roman; functions use lowercase Roman; functionals use capital Roman; constants use lowercase Greek or Roman; and quantities associated with a finite-dimensional Euclidean space (e.g., matrices and nodal vector spaces) use underbar.

Following is a description of the specific notation commonly used in this text. Since much of the development of concepts rests heavily on visualization, we have made several tacit assumptions about this notation that generally simplify the discussion, but are not otherwise essential. These assumptions are indicated in square brackets below; they are meant to hold unless otherwise indicated.

h	Generic mesh size [mesh sizes are equal in each coordinate direction]; used in superscripts and subscripts, but may be dropped when understood
$2h$	Coarse-grid mesh size
$\underline{h}$	$\underline{h} = \begin{pmatrix} h \\ 2h \end{pmatrix}$ (refers to the *composite* grid, formed as the union of global and local uniform grids of different mesh sizes)
Ω, Ω_h	(Open) regions in $\Re^d$ [regions are simply connected, $d = 2$]; $\overline{\Omega} =$ closure; $\partial\Omega = \overline{\Omega}\backslash\Omega =$ boundary; $C\Omega =$ complement
H, H_1, H_2, S^h, T^h	Continuum function spaces (including discrete finite element spaces)
$\text{Supp}(v), \text{Supp}(S)$	Support of the function v, union of the support of all functions in S
S_T	Maximal subspace of S of functions whose support is contained in $\text{Supp}(T)$
$S^{2h,h}, T^{2h,h}$	Subspaces common to coarse and fine levels: $S^{2h,h} = S^{2h} \cap S^h$, $T^{2h,h} = T^{2h} \cap T^h$
R, R^h	Objective functionals for variational problems
K, K^h	Operators for strongly formulated equations
k, k^h	Forms for weakly formulated equations
$\Omega^{2h}, \Omega^h, \Omega^{\underline{h}}$	Grids [Ω^{2h} and Ω^h are uniform; $\Omega^{\underline{h}}$ is the union of its uniform *subgrids* ($\Omega^{\underline{h}} = \Omega^{2h} \cup \Omega^h$); Ω^h is *aligned* with Ω^{2h}; grids "cover" their associated regions in the sense that the region is enclosed by the boundary of the grid]; grids exclude Dirichlet boundary points; *patch* = local rectangular uniform grid; *level* = uniform subgrid (i.e., union of patches of same mesh size)
P_k, P_{ij}	Grid points (nodes)
P_S	Orthogonal projection operator onto S

$w^h_{(ij)}$	*Hat* function in the finite element space S^h which has the value one at grid point P^h_{ij} and zero at all other grid points
$\langle \cdot, \cdot \rangle, \| \cdot \|$	$L_2(\Omega)$ or Euclidean inner product, norm
$\|\| \cdot \|\|$	*Energy norm*, used only for Poisson's equation: $\|\|u\|\| = \|\nabla u\|$
$\|\| \cdot \|\|_h$	*Discrete energy norm* associated with a self-adjoint positive-definite linear operator L^h: $\|\|u^h\|\| = \langle u^h, L^h u^h \rangle^{1/2}$
$H^1_0(\Omega)$	First-order Sobolev space of functions that satisfy (in the usual Sobolev sense) homogeneous boundary conditions: $H^1_0(\Omega) = \{u : \|u\|$ and $\|\nabla u\|$ defined, $u/\partial\Omega = 0\}$
I^h_{2h}, I^{2h}_h	Intergrid transfer matrices (*interpolation, restriction*)
$(\nu_1, \nu_2)V$	Multilevel V-*cycle* with ν_1 and ν_2 relaxations before and after coarsening, respectively
I	Identity operator (on continuum function spaces)
$\underline{I}$	Identity matrix
L^*	Operator adjoint
$\underline{L}^t$	Matrix transpose
∇	Divergence operator: $\nabla = \begin{pmatrix} \frac{\partial}{\partial x} \\ \frac{\partial}{\partial y} \end{pmatrix}$ in two dimensions
Δ	Laplacian operator: $\Delta = \nabla \cdot \nabla$
$\Re^n, \Re^h$	Euclidean space of dimension n, Euclidean space corresponding to the grid of mesh size h
τ^{2h}_h	Full approximation scheme *tau correction* term (cf. [Brandt 1984])
S^∞	Space spanned by elements of S

We will also use standard set notation, including $\emptyset$ (*the empty set*), $\forall$ (*for every*), $\in$ (*contained in*), $\subset$ (*is a subset of*), $\not\subset$ (*is not a subset of*), $\cap$ (*intersect*), and $\backslash$ (*intersect the complement of*).

To avoid iteration subscripts, approximations such as u are dynamic quantities that can change assignment in an algorithm by a statement of the form $u \leftarrow G(u)$. This is understood to mean that the new assignment of u is the result of applying G to the old one.

A superscript asterisk (*) on a vector is used to denote exact solutions and the symbol e is reserved to denote error. For example, $u^h - e^h = u^{h^*}$ represents the relationship between the approximation u^h and exact solution u^{h^*} on S^h. The various unknowns, errors, and error measures used in this monograph are as follows:

$$u^* \quad \text{Continuum solution (on } H)$$
$$u^{h^*} \quad \text{Discrete solution (on } S^h)$$
$$u^h \quad \text{Discrete approximation (in } S^h)$$
$$e \quad \textit{Actual error}: e = u^h - u^*$$
$$e^* \quad \textit{Discretization error}: e^* = u^{h^*} - u^*$$
$$e^h \quad \textit{Algebraic error}: e^h = u^h - u^{h^*}$$
$$r \quad \textit{Residual error}: r = K(u)$$
$$E_R, E_R^h, E_K, E_K^h \quad \text{Error measures: } E_R(e) = R(u^h) - R(u^*),$$
$$E_R^h(e^h) = R(u^h) - R(u^{h^*}),\ E_K(e) = \|K(u^h)\|,$$
$$E_K^h(e^h) = \|K^h(u^h)\|$$

We should point out that the term *multilevel* has typically been used in the open literature as a generalization of the term *multigrid*. The term multigrid usually refers specifically to methods that are applied to the solution of PDEs discretized on a given (fine) grid and that use global relaxation schemes and global coarse grids for corrections. Our use of the term multilevel refers to a slight generalization of this type of method, which includes the case that the discretization uses a composite grid (e.g., consisting of uniform coarse and fine grids; see Figure 1.1) and relaxation on one or more levels is restricted to the local uniform grids.

Acronyms used in this monograph are included below:

AFAC	Asynchronous FAC
AMG	Algebraic multigrid method
BLAS	Basic linear algebra subroutines (cf. [Dongarra et al. 1979]).
FAC	Fast adaptive composite grid method
FAS	Full approximation scheme
FMG	Full multigrid method
FVE	Finite volume element method
GML	Multilevel Galerkin or Petrov-Galerkin method
MG	Multigrid method
PDE	Partial differential equation
PML	Multilevel projection method
RML	Multilevel Rayleigh-Ritz method
UG	Unigrid
WU	Work unit

Algorithms used here include:

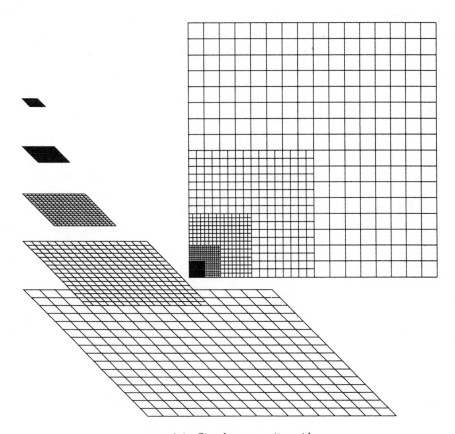

FIG. 1.1. *Simple composite grid.*

AFAC^h	Asynchronous FAC
C^h	Coarse-grid correction
D^h	Directional iteration
FAC^h	Fast adaptive composite grid method
G^h	Relaxation (block Gauss-Seidel)
GML^h	Multilevel Galerkin or Petrov-Galerkin method
MG^h	Multigrid method
PML^h	Multilevel projection method
RML^h	Multilevel Rayleigh-Ritz method

1.3. Prototype problems.

To provide a concrete understanding of the concepts developed here, we will illustrate them in terms of two simple prototypes, a linear equation and an eigenvalue problem, both involving the Laplacian.

Let H_1 and H_2 be appropriate Hilbert spaces of function defined on the unit square, $\Omega = [0,1] \times [0,1]$. For simplicity, we assume that each function, u, in H_1

in some sense satisfies the homogeneous Dirichlet condition $u(z) = 0$, $z \in \partial\Omega$. Let $L : H_1 \to H_2$ be defined by $Lu = -\Delta u$, $u \in H_1$, and suppose that $f \in H_2$ is given. Then the prototypes we consider are given by

$$(1.1) \qquad\qquad Lu = f, \quad u \in H_1,$$

and

$$(1.2) \qquad\qquad \langle u, u \rangle Lu = \langle u, Lu \rangle u, \quad u \in H_1.$$

Here, $\langle \cdot, \cdot \rangle$ is the L_2 inner product on H_1. (We have chosen to write the eigenvalue problem using a form that is defined on all of H_1, including the origin. This simplifies treatment of both (1.1) and (1.2). However, it should be kept in mind that we are in fact seeking *some nonzero* $u \in H_1$ that satisfies (1.2) and for which the eigenvalue $\lambda = \frac{\langle u, Lu \rangle}{\langle u, u \rangle}$ is *minimal*.) To recast these equations in the forms suitable for projection methods, define $K_L : H_1 \to H_2$ and $K_E : H_1 \to H_2$ as follows:

$$(1.3) \qquad\qquad K_L(u) = Lu - f$$

and

$$(1.4) \qquad\qquad K_E(u) = \langle u, u \rangle Lu - \langle u, Lu \rangle u.$$

Then, with $K(u)$ denoting either $K_L(u)$ or $K_E(u)$, our two prototypes can be written collectively as the *equation*

$$(1.5) \qquad\qquad K(u) = 0, \quad u \in H_1.$$

We will make use of the weak form of (1.5), which is developed as follows. Define the forms $k_L : H_2 \times H_1 \to \Re$ and $k_E : H_2 \times H_1 \to \Re$ by

$$(1.6) \qquad\qquad k_L(v, u) = \langle \nabla v, \nabla u \rangle - \langle v, f \rangle$$

and

$$(1.7) \qquad\qquad k_E(v, u) = \langle u, u \rangle \langle \nabla v, \nabla u \rangle - \langle \nabla u, \nabla u \rangle \langle v, u \rangle.$$

With $k(v, u)$ denoting either $k_L(v, u)$ or $k_E(v, u)$, then corresponding to (1.5) is the *weak form*

$$(1.8) \qquad\qquad k(v, u) = 0, \quad u \in H_1, \quad \forall v \in H_2.$$

Natural choices for the function spaces associated with k_L and k_E are $H_1 = H_2 = H \equiv H_0^1(\Omega)$, the first-order Sobolev space of functions on Ω satisfying the homogeneous Dirichlet boundary conditions. In this case, we have the *weak form*

(1.9) $k(v, u) = 0, \quad u \in H, \quad \forall v \in H.$

A major advantage of the weak forms (1.8) and (1.9) over the strong form (1.5) is that their admissible function spaces (i.e., H_1 and H, respectively) may generally be larger. For example, the form k_L requires only first-order differentiability of u in the L_2 sense, while K_L requires second-order differentiability presumably everywhere. This difference is essential to the discretization process: k_L admits continuous piecewise linear functions, while K_L does not in the classical sense.

We will have occasion in this monograph to discuss (1.5), (1.8), and (1.9) collectively. In order to avoid the inconvenience of including both the strong and weak forms, we will instead refer to (1.5) with the understanding that its meaning may be in the weak sense. Specifically, whenever we refer to (1.5), it should be understood that the equality may not be meant in the classical sense: whenever the stronger form is not strictly defined, $K(u) = 0$ will instead be taken to mean $\langle v, K(u) \rangle = 0$ for all $v \in H_2$ or, in the transformed sense, $k(v, u) = 0$ for all $v \in H_2$. This is to be understood for (1.5) and any equation derived from it. For example, if P and Q are projection operators, then by $PKQu = 0$ we may mean $k(Pv, Qu) = 0$ for all $v \in H_2$. The definition of the operator PKQ itself may be meant in this context, that is, $PKQu$ is the element of PH_2 satisfying

$$\langle v, PKQu \rangle = k(Pv, Qu) \quad \forall v \in H_2.$$

To recast these equations in their variational form, consider the functionals $\ell : H_1 \to \Re$ and $m : H_1 \to \Re$ given by

$$\ell(u) = \|\nabla u\|^2$$

and

$$m(u) = \|u\|^2,$$

where $\|\cdot\|$ is the L_2 norm on H_1. Assume that $f \in H_2$ and define the functionals $R_L : H_1 \to \Re$ (*energy*) and $R_E : H_1 \setminus \{0\} \to \Re$ (*Rayleigh quotient*) as follows:

(1.10) $R_L(u) = \ell(u) - 2\langle u, f \rangle$

and

(1.11) $R_E(u) = \dfrac{\ell(u)}{m(u)}.$

Note that, for appropriate u, $K_L(u)$ and $K_E(u)$ are proportional to the respective gradients of $R_L(u)$ and $R_E(u)$. Then, with $R(u)$ denoting either $R_L(u)$ or

$R_E(u)$ and H denoting H_1, our prototype variational problems can be written collectively as the *variation*

$$(1.12) \qquad\qquad R(u) = \min_{v \in H} R(v), \quad u \in H.$$

For the eigenvalue case in (1.12), technically we should replace H by $H \setminus \{0\}$. We assume this to be understood. Also for this case, the variational form amounts to specifying that we are targeting only the smallest eigenvalue of $L = -\Delta$. For definiteness, we will henceforth add this constraint implicitly to (1.5), (1.8), and (1.9). For example, with $K = K_E$, we take (1.5) to mean: "Find a nonzero solution of $K_E(u) = 0$ with smallest $\lambda = \frac{\langle u, Lu \rangle}{\langle u, u \rangle}$."

There are several important properties of the prototypes that make the general problems suitable for the numerical discretization and solution methods we are about to develop. For example, all formulations for the prototypes have certain ellipticity properties that generally guarantee unique solvability (up to a scale factor for the eigenproblem). In order to maintain more generality, and because we are not concerned here with theoretical issues, no assumptions of this kind will be introduced for the general formulations, (1.5), (1.9), or (1.12). However, it should be understood that some of these properties (e.g., that R *has* a minimum) are necessary for our discussion to make any sense.

An interesting way to generalize (1.12) would be to add constraints. In fact, the eigenproblem already has the implicit restriction $u \neq 0$, although this simple constraint has little impact on the numerical process we will develop. In general, however, the introduction of constraints leads to greater complexity in the number and type of numerical solution methods. We therefore restrict this monograph to the unconstrained case. (For recent work in the direction of projection methods for constrained optimization, see [Gelman and Mandel 1990].)

1.4. Discretization by projections.

One of the basic principles of multilevel methodology is to design the coarse-grid correction process in a way that is compatible with the discretization method. This is not a firm principle: there are instances where it is advantageous to use a coarsening technique that is in some ways superior to the discretization scheme, particularly when the latter is inaccurate; consider the double discretization methods in [Brandt 1984, pages 103–106]; also, the algebraic multigrid method (AMG) is based on a variational-type coarsening, but applied to matrices that do not necessarily arise from this type of discretization. However, this coarsening principle does provide a general multilevel design criterion that can lead to very effective algorithms. In any case, it is natural to relate two discrete levels guided by the discretization itself.

To be more specific, but speaking loosely, suppose we are given a general discretization procedure, which relates the PDE (in weak or strong form) to a given discrete problem on level h (either finite element or nodal vector spaces).

By this is meant that we are given a mapping (possibly a projection) from the PDE space, H, to the level h space, S^h, and, just as important, that we also are given a mapping from S^h to H. Write these mappings as $Q^h : H \to S^h$ and $Q_h : S^h \to H$, and suppose that we have analogous mappings for the subspace S^{2h} of S^h. Then the relationship between levels $2h$ and h is defined by constructing interlevel transfers for which the diagram in Figure 1.2 commutes. That is, we define $Q_h^{2h} : S^h \to S^{2h}$ and $Q_{2h}^h : S^{2h} \to S^h$ implicitly by $Q^{2h} = Q_h^{2h} \cdot Q^h$ and $Q_{2h} = Q_h \cdot Q_{2h}^h$. Of course, we must be sure that these definitions are realistic, that is, that these interlevel transfer operators are easily implemented. For the conforming projection methods considered here, this is certainly true as we shall see; in fact, this is one of the major motives for our focus on projection techniques.

In addition to determining interlevel transfers, the problem must be specified on each level. The attempt to be compatible with the procedures used to define the fine-level problem leads us to choose the level h problem simply as the one that is induced by the discretization. This is the approach taken in this monograph. As we shall see, the form that the coarse-level problem takes is often a generalization of the fine-grid form precisely because the coarse-level unknown is an approximation to the fine-level error, not its solution; that is, the coarse-level correction is of the form $u^h + u^{2h}$, and the presence of $u^h \neq 0$ creates a term in the coarse-level problem that is not generally present on the fine level (unless the problem is a linear equation or quadratic variation).

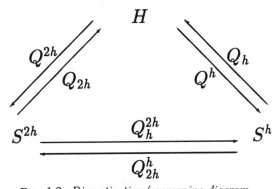

FIG. 1.2. *Discretization/coarsening diagram.*

One of the main purposes of the foregoing discussion is to emphasize that, for projection discretization methods, taking this compatibility approach actually dictates the major coarsening components, which consist of the intergrid transfer operators and coarse-grid problems. We will see that this approach also guides the choice of relaxation. To be sure, other multilevel components must be determined, including the type and schedule of cycling, the possible use and placement of local grids, and the method of evaluating integrals and other quantities that must be approximated. The proper choice of these other components is no doubt critical to performance. But the major effort in applying multilevel methods is almost always devoted to the correct choice of coarsening and relaxation. Thus, being guided by the projection scheme substantially narrows the focus of decision making. This does not mean that we have trivialized the task of multilevel design; in fact, proper choice of the underlying subspaces for relaxation and coarsening may be far from obvious. Nor have we fully committed ourselves to the projection approach; approximations to it, and even wholly new types of coarsening, may be better in some circumstances. However, it is very useful to begin the task of multilevel design with a formalism to guide the major choices. The task could then focus on determining the relaxation and coarsening subspaces that work together quickly to eliminate all error components. We will illustrate the results of several such design tasks in Chapter 4.

This somewhat lengthy motivation brings us up to the description of projection discretization methods. Starting first with the *equation*, (1.5), let $S^h \subset H_1$ and $T^h \subset H_2$ be finite-dimensional subspaces (possibly of different dimension) and let $P_{S^h} : H_1 \to S^h$ and $P_{T^h} : H_2 \to T^h$ be orthogonal projections onto S^h and T^h, respectively. Then the projection discretization of (1.5) is given by the finite-dimensional problem

$$(1.13) \qquad P_{T^h} K(P_{S^h} u) = 0, \quad u \in H_1.$$

Let $K^h : S^h \to T^h$ be given by $K^h(u^h) \equiv P_{T^h} K(P_{S^h} u^h) = P_{T^h} K(u^h)$, $u^h \in S^h$. For example, assuming the case $T^h = S^h$, then $K_L^h(u^h) = L^h u^h - f^h$, where $L^h = P_{S^h} L P_{S^h}$ and $f^h = P_{S^h} f$. Remember that the definition of L^h may be meant in the weak sense. Similarly, $K_E^h(u^h) = \langle u^h, M^h u^h \rangle L^h u^h - \langle u^h, L^h u^h \rangle M^h u^h$, where $L^h = P_{S^h} L P_{S^h}$ and $M^h = P_{S^h}$. Then (1.13) can be rewritten as

$$(1.14) \qquad K^h(u^h) = 0, \quad u^h \in S^h.$$

Note that the weak interpretation of (1.13) or (1.14) is

$$(1.15) \qquad k(P_{T^h} v, P_{S^h} u) = 0, \quad u \in H_1, \quad \forall v \in H_2.$$

Defining $k^h(v^h, u^h) \equiv k(P_{T^h} v^h, P_{S^h} u^h) = k(v^h, u^h)$, $u^h \in S^h$, $v^h \in T^h$, then this becomes

$$(1.16) \qquad k^h(v^h, u^h) = 0, \quad u^h \in S^h, \quad \forall v^h \in T^h.$$

We refer to any of these discrete formulations, (1.13)–(1.16), as the *Petrov-Galerkin* form. In the special case $S^h = T^h$, they will be referred to as the *Galerkin* form.

The projection discretization of the *variation*, (1.12), is simply

$$(1.17) \qquad R(P_{S^h} u) = \min_{v \in H} R(P_{S^h} v), \quad u \in H.$$

Let $R^h(u^h) \equiv R(P_{S^h} u^h) = R(u^h)$, $u^h \in S^h$. For example, $R_L^h(u^h) = \langle u^h, L^h u^h \rangle - 2\langle u^h, f^h \rangle$, where L^h and f^h are defined as above. Similarly, $R_E^h(u^h) = \frac{\langle u^h, L^h u^h \rangle}{\langle u^h, M^h u^h \rangle}$. Then (1.17) can be rewritten as

$$(1.18) \qquad R^h(u^h) = \min_{v^h \in S^h} R^h(v^h), \quad u^h \in S^h.$$

We refer to either (1.17) or (1.18) as the *Rayleigh-Ritz* discretization of (1.12).

The notation we have used thus far is suitable for discretization on a given global uniform grid. Since a major emphasis of this monograph is on local refinement methods, we now turn to the development of the projection discretizations suitable for composite grids. To this end, let S^h and S^{2h} be nontrivial finite-dimensional subspaces of H_1 and let T^h and T^{2h} be nontrivial finite-dimensional subspaces of H_2. We suggest by this notation that S^h and T^h correspond to global *uniform* grids of mesh size h, and that S^{2h} and T^{2h} correspond to their uniform *subgrids* of mesh size $2h$, because this situation is typical of many multigrid implementations. However, none of these properties are really necessary for our discussion: none of the levels need be uniform, the mesh ratios can be different from 2, the coarse levels need not be subgrids of the fine level, and the fine grids may be local. Nevertheless, for simplicity of development, we will henceforth assume that the subspaces are *conforming*, which for the case that the fine grid is global is characterized by the assumption that

$$(1.19) \qquad\qquad S^{2h} \subset S^h \quad \text{and} \quad T^{2h} \subset T^h.$$

Consider the case where the fine-grid regions $\mathrm{Supp}(S^h)$ and $\mathrm{Supp}(T^h)$ may be proper subregions of Ω. Here, $\mathrm{Supp}(S) = \bigcup_{v \in S} \mathrm{Supp}(v)$, where $\mathrm{Supp}(v)$ is the maximal subregion of Ω on which v is nonzero (i.e., the *support* of v). Let $S^{2h}_{S^h}$ and $T^{2h}_{T^h}$ denote the subspaces of S^{2h} and T^{2h} of functions whose support are contained in $\mathrm{Supp}(S^h)$ and $\mathrm{Supp}(T^h)$, respectively. Then by *conforming* in this general case we mean that

$$(1.20)$$
$$\mathrm{Supp}(S^{2h}_{S^h}) = \mathrm{Supp}(S^h), \ \mathrm{Supp}(T^{2h}_{T^h}) = \mathrm{Supp}(T^h), \ S^{2h}_{S^h} \subset S^h, \ \text{and} \ T^{2h}_{T^h} \subset T^h.$$

The first two of these rather technical relations ensure that the grid interfaces for the local regions $\mathrm{Supp}(S^h)$ and $\mathrm{Supp}(T^h)$ are actually coarse-grid lines. Note that (1.19) is just a special case of (1.20).

Now define the *composite* grid spaces by

$$(1.21) \qquad\qquad S^{\underline{h}} = S^{2h} + S^h \quad \text{and} \quad T^{\underline{h}} = T^{2h} + T^h.$$

Here we use underbar on h for denoting the composite grid vector of subgrid mesh sizes,

$$\underline{h} = \begin{pmatrix} h \\ 2h \end{pmatrix}.$$

Note that $S^{\underline{h}} = S^h$ and $T^{\underline{h}} = T^h$ in the global grid case because we are assuming the conforming property, (1.19). Note also that, in general, the sums in (1.21) are not direct because $S^{2h} \cap S^h \neq \emptyset$ and $T^{2h} \cap T^h \neq \emptyset$. (In fact, in the global grid case, $S^{2h} \cap S^h = S^{2h}$ and $T^{2h} \cap T^h = T^{2h}$.) This observation is important to understanding the limits to parallelizability of the *fast adaptive composite grid method* (FAC) and the motive for the modifications to it that lead to *asynchronous FAC* (AFAC). We will briefly discuss these points in Chapter 2, §2.7. (See [McCormick 1989, pages 129–133] for a more thorough discussion.)

The composite grid problem that we wish to solve is either the equation, (1.14), or the variation, (1.18), but posed now on the composite grid space, $S^{\underline{h}}$ and $T^{\underline{h}}$. In particular, letting $K^{\underline{h}}(u^{\underline{h}}) \equiv P_{T^{\underline{h}}} K(P_{S^{\underline{h}}} u^{\underline{h}})$, $u^{\underline{h}} \in S^{\underline{h}}$, and $R^{\underline{h}}(u^{\underline{h}}) \equiv R(P_{S^{\underline{h}}} u^{\underline{h}})$, $u^{\underline{h}} \in S^{\underline{h}}$, then two forms of the composite grid problem are the equation

$$(1.22) \qquad\qquad K^{\underline{h}}(u^{\underline{h}}) = 0, \quad u^{\underline{h}} \in S^{\underline{h}},$$

and the variation

$$(1.23) \qquad\qquad R^{\underline{h}}(u^{\underline{h}}) = \min_{v^{\underline{h}} \in S^{\underline{h}}} R^{\underline{h}}(v^{\underline{h}}), \quad u^{\underline{h}} \in S^{\underline{h}}.$$

Of course, (1.22) may be taken in the weak sense

$$(1.24) \qquad\qquad k^{\underline{h}}(v^{\underline{h}}, u^{\underline{h}}) = 0, \quad u^{\underline{h}} \in S^{\underline{h}}, \quad \forall v^{\underline{h}} \in T^{\underline{h}},$$

where $k^{\underline{h}}(v^{\underline{h}}, u^{\underline{h}}) \equiv k(P_{T^{\underline{h}}} v^{\underline{h}}, P_{S^{\underline{h}}} u^{\underline{h}})$, $u^{\underline{h}} \in S^{\underline{h}}$, $v^{\underline{h}} \in T^{\underline{h}}$.

1.5. Realizability and nodal representations.

Both the Petrov-Galerkin and Rayleigh-Ritz discretization methods are deceptively simple, primarily because we are dealing here with projections and subspaces. We have not yet faced the issue of how these discretizations are to be represented in a computational environment, which can be rather involved. The usual approach is to specify the subspaces S^h and T^h by finite elements, choose a locally supported basis for each, then rewrite the discrete problem in terms of the coefficients of the unknown u^h expanded in the basis for S^h.

For example, for our prototype problems in the form of (1.5) or (1.12) on global grids, we may choose $S^h = T^h$ as the subspace of functions in H_1 that are piecewise bilinear with respect to the cells of a uniform grid, Ω^h, on Ω. For the basis, we would choose the usual *hat* functions, which are defined as follows. Let the global grid Ω^h have interior nodes P_{pq}, $1 \le p, q \le n$. Then the hat function associated with P_{pq} is the element $w^h_{(pq)} \in S^h$ that satisfies

$$w^h_{(pq)}(P) = \begin{cases} 1 & P = P_{pq} \\ 0 & P \in \Omega^h \setminus \{P_{pq}\}. \end{cases}$$

The computational problem associated with (1.14) is to find the coefficients, u^h_{pq}, in the expression

$$(1.25) \qquad\qquad u^h = \sum_{p,q=1}^{n} u^h_{pq} w^h_{(pq)}$$

such that

$$(1.26) \qquad k^h \left(w^h_{(ij)}, \sum_{p,q=1}^{n} u^h_{pq} w^h_{(pq)} \right) = 0 \qquad \forall\, i, j \in \{1, 2, \cdots, n\}.$$

(Here we must consider the weak form, (1.16), of (1.14) because $K^h(w^h_{(pq)})$ is not defined in the classical sense. Remember that $T^h = S^h$ for both prototypes.) Let $\Re^h$ denote the Euclidean space of dimension n^2 corresponding to the nodal vectors for grid Ω^h. Then, for the linear equation prototype, (1.1), this computational problem can be written as

$$(1.27) \qquad \underline{L}^h \underline{u}^h = \underline{f}^h, \quad \underline{u}^h \in \Re^h.$$

Here, $\underline{L}^h$ is the matrix whose $(ij) - (pq)$ entry (i.e., entry in the row associated with P_{ij} and column associated with P_{pq}) is given by

$$(1.28) \qquad (\underline{L}^h)_{ij-pq} = \langle \nabla w^h_{(ij)}, \nabla w^h_{(pq)} \rangle = \begin{cases} \frac{8}{3} & p = i,\ q = j \\ -\frac{1}{3} & \max\{|p - i|, |q - j|\} = 1 \\ 0 & \text{otherwise,} \end{cases}$$

$\underline{u}^h$ is the vector of nodal values u^h_{ij}, and $\underline{f}^h$ is the vector with entries

$$(1.29) \qquad f^h_{ij} \equiv \int_0^1 \int_0^1 f(x, y) w^h_{ij}(x, y)\, dx dy.$$

We used the bilinearity property of the form k for (1.1) to *realize* the equation (1.27), meaning that this property of k allows us to reduce (1.14) to an equation that is computationally expressible in the sense that all of the operations defining it are elementary. We call (1.14) *realizable* in this case, which is an important concept because it determines whether we can really begin to *compute* discrete approximations. Actually, we have not quite reduced problem (1.21) to elementary operations because integrals are involved in (1.29). The exact evaluation of $\underline{f}^h$ is generally not possible, so in a strict sense (1.14) is not realizable. However, this is a classical situation to treat in practice: we can, for example, simply *assume* f has some property (e.g., $f \in S^h$) that allows exact evaluation of the integral in (1.29), then use standard Sobolev space arguments to determine the error introduced by this assumption. (See [Ciarlet 1978, §4.1], for example.) To simplify discussion, and because discretization error does not directly concern us here anyway, we will make such an assumption throughout the remainder of this monograph. More precisely, we will henceforth assume that the discretization of any *given* function by formulas like (1.29) is done exactly. In fact, we also make similar implicit assumptions on other subprincipal terms in the PDE (e.g., coefficients) so that they do not by themselves prevent realizability of the discretization.

To see that (1.14) is realizable for the eigenvalue problem, (1.2), we again use the basis expansion, (1.25), and consider the case $k = k_E$ in (1.26). A simple calculation using the bilinearity of the inner products defining k_E yields

$$(1.30) \qquad \langle \underline{u}^h, \underline{M}^h \underline{u}^h \rangle \underline{L}^h \underline{u}^h = \langle \underline{u}^h, \underline{L}^h \underline{u}^h \rangle \underline{M}^h \underline{u}^h, \quad \underline{u}^h \in \Re^h,$$

where $\underline{L}^h$ and $\underline{u}^h$ are defined as above, $\underline{M}^h$ is the matrix whose $(ij) - (pq)$ entry is given by

$$(1.31) \qquad (\underline{M}^h)_{ij-pq} = \langle w^h_{(ij)}, w^h_{(pq)} \rangle = \begin{cases} \frac{16}{36} h^2 & p = i,\ q = j \\ \frac{4}{36} h^2 & |p - i| + |q - j| = 1 \\ \frac{1}{36} h^2 & |p - i| = |q - j| = 1 \\ 0 & \text{otherwise,} \end{cases}$$

and $\langle \cdot, \cdot \rangle$ denotes the usual Euclidean inner product on $\Re^h$.

Computational forms for the discrete variation, (1.18), for both the linear and eigenvalue prototypes are easily obtained. This is done by way of the expression (1.25), which is used to define $\underline{R}^h(\underline{u}^h) = R^h \left(\sum_{p,q=1}^n u^h_{pq} w^h_{(pq)} \right)$, $\underline{u}^h \in \Re^h$. For the linear prototype, (1.1), we have

$$(1.32) \qquad \underline{R}^h(\underline{u}^h) = \langle \underline{u}^h, \underline{L}^h \underline{u}^h \rangle - 2 \langle \underline{u}^h, \underline{f}^h \rangle, \quad \underline{u}^h \in \Re^h.$$

For the eigenvalue prototype, (1.2), we have

$$(1.33) \qquad \underline{R}^h(\underline{u}^h) = \frac{\langle \underline{u}^h, \underline{L}^h \underline{u}^h \rangle}{\langle \underline{u}^h, \underline{M}^h \underline{u}^h \rangle}, \quad \underline{u}^h \in \Re^h \setminus \{\underline{0}\}.$$

These are just the standard variational forms for the matrix problems (1.27) and (1.30), respectively. In fact, (1.27) and (1.30) can be derived by setting to zero the gradients of the functionals defined in (1.32) and (1.33), respectively. These forms show that the variational discretizations of both our prototypes are realizable.

Realizability does not hold in general, but there appear to be many important practical cases where it does, as illustrated in Chapter 4. In other practical applications, computable approximations to the coarsening strategy may be suitable. In any event, this concept will become more relevant in the next chapter, where we consider multilevel coarsening strategies based on projections. The issue then is realizability of the coarse-level problem, which is posed to approximate the fine-level error. This is different from realizability of the discretization, which is posed to approximate the continuum solution. However, it should become apparent that the two are very closely related, and that theory developed to ensure (approximate) coarse-level realizability will no doubt rest on an understanding of discretization realizability.

To illustrate the procedure for constructing composite grid representations, consider the linear equation prototype, (1.1). The discretization of the weak form, (1.16), produces a matrix $\underline{L}^h$ that has essentially the same entries as $\underline{L}^h$ given in (1.28) associated with the uniform regions of the *composite grid* $\Omega^h = \Omega^{2h} \cup \Omega^h$, i.e., at points of $\Omega^h \setminus \Gamma^h$, where $\Gamma^h \equiv \Omega^{2h} \cap \partial\Omega^h$ is the composite grid *interface*. The only special entries of $\underline{L}^h$ are those associated with the points of Γ^h.

To see how these interface stencils can be generated, consider Figure 1.3, which depicts a typical point, P_{ij}, in Γ^h together with its composite grid neighbors. (Here we index points according to the fine-grid mesh size; for example, in Figure 1.3, $P_{i+1\,j+1}$ corresponds to $P_{ij} + (h,h)$.) Most of the coefficients of $\underline{L}^h$ corresponding to its ijth row can be determined by using symmetry of $\underline{L}^h$ and knowledge of the coefficients in rows corresponding to the uniform region of Ω^h. For example, the coefficient corresponding to column $i-2\,j$ is $-\frac{1}{3}$. As a more complicated example, consider row $i+1\,j$ of $\underline{L}^h$. It would seem to connect point $P_{i+1\,j}$ to points $P_{i\,j\pm1}$ each with $-\frac{1}{3}$. But $P_{i\,j\pm1}$ are *slave points* (i.e., their values are not free but are rather determined by neighboring values), so use of bilinear interpolation means that the node values there are defined to be the average of the node values at the interface neighbors (e.g., $\underline{u}^h(P_{i\,j+1}) = (\underline{u}^h(P_{ij}) + \underline{u}^h(P_{i\,j+2}))/2$). Thus, there are no columns labeled $i\,j\pm1$ in L^h. Instead, in row $i+1\,j$ of $\underline{L}^h$, the slave point $P_{i\,j+1}$ creates the additional coefficient $-\frac{1}{6}$ in both columns ij and $i\,j+2$. Thus, the coefficient in row $i+1\,j$ associated with P_{ij} is $-\frac{1}{3} - \frac{1}{6} - \frac{1}{6} = -\frac{2}{3}$. Similarly, we can argue that the coefficient associated with P_{ij} in row $i+1\,j+1$ is $-\frac{1}{3} - \frac{1}{6}$, where the second term comes from the coefficient of the slave point $P_{i\,j+1}$, which is just $-\frac{1}{2}$. Continuing to row $i+1\,j+2$, we see that the coefficient associated with P_{ij} is $-\frac{1}{6}$ (via slave point $P_{i\,j+1}$). Hence, by symmetry of $\underline{L}^h$, the coefficients in row ij associated with points $P_{i+1\,j}$, $P_{i+1\,j+1}$, $P_{i+1\,j+2}$, $P_{i-1\,j}$, and $P_{i-1\,j+2}$ are $-\frac{2}{3}$, $-\frac{1}{2}$, $-\frac{1}{6}$, $-\frac{1}{3}$, and $-\frac{1}{3}$, respectively. This leaves determination of the coefficients associated with $P_{i\,j+2}$, which can be made directly; a simple calculation of $k_L^h(w_{(ij)}^h, w_{(i\,j+2)}^h)$ shows that $P_{i\,j+2} = 0$. Finally, the diagonal term must be such that the row sum is zero (because S^h contains functions that are constant in a region surrounding P_{ij} and because L applied to such functions must be zero inside this region). This implies that the entries for $\underline{L}^h$ in row ij associated with point P_{pq} are given by

$$
\left(\underline{L}^h\right)_{ij-pq} =
\begin{cases}
3 & p=i,\ q=j \\
-\frac{1}{3} & p=i-2,\ q=j,\ j\pm2 \\
-\frac{2}{3} & p=i+1,\ q=j \\
-\frac{1}{2} & p=i+1,\ q=j\pm1 \\
-\frac{1}{6} & p=i+1,\ q=j\pm2 \\
0 & \text{otherwise.}
\end{cases}
$$

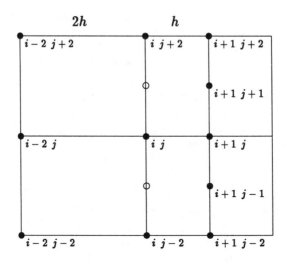

FIG. 1.3. *Composite grid interface points (○ denotes slave points).*

1.6.   Interlevel transfer matrices.

In a few instances we will illustrate basic principles by way of nodal vector representations for the prototype problems. For this purpose, we will need to use notation for the relationships between the various levels, Ω^{2h}, Ω^h, and $\Omega^{\underline{h}}$. Consider first the global-grid case, $\Omega^{2h} \subset \Omega^h$, and let $\Re^{2h}$ and $\Re^h$ denote the respective Euclidean spaces of coarse-grid and fine-grid nodal vectors. Then, because we are using bilinear interpolation for the prototypes, the relationship between grid $2h$ and grid h is given by the matrix $\underline{I}^h_{2h} : \Re^{2h} \rightarrow \Re^h$, which is defined as follows:

$\underline{I}^h_{2h} : \underline{u}^{2h} \mapsto \underline{u}^h$, where $\underline{u}^h$ is the nodal vector corresponding to $u^h \in S^h$, which is defined for $P \in \Omega^h$ by

$$u^h(P) = \begin{cases} u^{2h}(P) & P \in \Omega^{2h} \\ (u^{2h}(P - P_0) + u^{2h}(P + P_0))/2 & P \pm P_0 \in \Omega^{2h}, \\ & P_0 = P_x \text{ or } P_y \\ (u^h(P - P_x) + u^h(P + P_x))/2 & P \pm P_x \pm P_y \in \Omega^{2h}. \end{cases}$$

Here, $u^{2h} \in S^{2h}$ is the function with corresponding nodal vector $\underline{u}^{2h}$, $P_x = (h, 0)$, and $P_y = (0, h)$. (By $P \pm P_0 \in \Omega^{2h}$, we mean either $P + P_0 \in \Omega^{2h}$ or $P - P_0 \in \Omega^{2h}$, or both; by $P \pm P_x \pm P_y \in \Omega^{2h}$, we mean at least one of the possible four points is in Ω^{2h}. This rather complicated notation is due largely to the need to account for boundary points.)

The relationship from Ω^h to Ω^{2h} is given by the adjoint mapping,

(1.34) $$\underline{I}^{2h}_h = \left(\underline{I}^h_{2h} \right)^t,$$

which is proportional to the so-called *full-weighting* operator (cf. [Stüben and Trottenberg 1982, pages 14–16, 27]). It is easily verified that the definition of $\underline{L}^h$ in (1.27) satisfies

$$(1.35) \qquad\qquad \underline{L}^{2h} = I_h^{2h} \underline{L}^h I_{2h}^h.$$

Equation (1.35) is referred to as the *Galerkin condition*; (1.34) and (1.35) together are called the *variational conditions*. (See [McCormick 1987, pages 132–134].)

For the local refinement case, where $\Omega^{2h} \not\subset \Omega^h$, we need to define transfer matrices between levels Ω^{2h} and $\Omega^{\underline{h}}$ and between levels Ω^h and $\Omega^{\underline{h}}$. (Transfers between Ω^{2h} and Ω^h will not be needed because of the way we pose the basic methods: corrections from either of these grids are collected in the grid $\underline{h}$ approximation, so these grids need not communicate directly between themselves.) Let $\Re^{\underline{h}}$ denote the Euclidean space of composite grid nodal vectors. The definition of $\underline{I}_{2h}^{\underline{h}} : \Re^{2h} \to \Re^{\underline{h}}$ is analogous to the definition of $\underline{I}_{2h}^h$:

$\underline{I}_{2h}^{\underline{h}} : \underline{u}^{2h} \mapsto \underline{u}^{\underline{h}}$, where $\underline{u}^{\underline{h}}$ is the nodal vector corresponding to $u^{\underline{h}} \in S^{\underline{h}}$, which is defined for $P \in \Omega^{\underline{h}}$ by

$$u^{\underline{h}}(P) = \begin{cases} u^{2h}(P) & P \in \Omega^{2h} \\ (u^{2h}(P - P_0) + u^{2h}(P + P_0))/2 & P \pm P_0 \in \Omega^{2h}, \\ & P_0 = P_x \text{ or } P_y \\ (u^{\underline{h}}(P - P_x) + u^{\underline{h}}(P + P_x))/2 & P \pm P_x \pm P_y \in \Omega^{2h}. \end{cases}$$

Here we should emphasize that our construction of the finite element spaces implies that the *interface* between the coarse-grid $(2h)$ and fine-grid (h) regions in $\Omega^{\underline{h}}$ contains only coarse-grid points. Specifically, this interface, $\Gamma^{\underline{h}} \equiv \Omega^{2h} \cap \partial\Omega^h$, is defined as the set of coarse-grid points on the *internal boundary*, which is the boundary of the region covered by the local grid, Ω^h, *excluding* the real boundary $\partial\Omega$. Let P be *any* point on the internal boundary. The boundary values for S^h are assumed to be zero, so $u^h \in S^h$ implies that $u^h(P) = 0$. Since each $u^{\underline{h}} \in S^{\underline{h}}$ can be written as the sum $u^{\underline{h}} = u^{2h} + u^h$ for some $u^{2h} \in S^{2h}$ and $u^h \in S^h$, then $u^{\underline{h}}(P) = u^{2h}(P)$. In other words, the internal boundary is a coarse-grid line. This means that the midpoint between two neighboring coarse-grid points in $\Gamma^{\underline{h}}$ is treated in practice as a *slave point*: the value of $u^{\underline{h}}$ there is imposed to be the average of $u^{\underline{h}}$ at these coarse-grid neighbors. This constraint is automatically incorporated into the definition of $\underline{I}_{2h}^{\underline{h}}$ given above.

The definition of the matrix $\underline{I}_h^{\underline{h}} : \Re^h \to \Re^{\underline{h}}$ is given simply as follows:

$\underline{I}_h^{\underline{h}} : \underline{u}^h \to \underline{u}^{\underline{h}}$, where $\underline{u}^{\underline{h}}$ is the nodal vector corresponding to $u^{\underline{h}} \in S^{\underline{h}}$, which is defined for $P \in \Omega^{\underline{h}}$ by

$$u^{\underline{h}}(P) = \begin{cases} u^h(P) & P \in \Omega^h \\ 0 & \text{otherwise.} \end{cases}$$

This represents an imbedding of $\Re^h$ into $\Re^{\underline{h}}$.

The restriction operators $\underline{I}^{2h}_h : \Re^{\underline{h}} \to \Re^{2h}$ and $\underline{I}^h_{\underline{h}} : \Re^{\underline{h}} \to \Re^h$ are defined as adjoints of interpolation:

$$\underline{I}^{2h}_h = \left(\underline{I}^h_{2h} \right)^t$$

and

$$\underline{I}^h_{\underline{h}} = \left(\underline{I}^h_h \right)^t.$$

At the end of the last section, we illustrated how the composite grid matrix $\underline{L}^{\underline{h}}$ associated with the linear prototype, (1.1), may be constructed. It is not difficult to argue from basic principles, or from examining in detail the matrix entries, that all of the grid operators satisfy the Galerkin principle. For example,

$$\underline{L}^h = \underline{I}^h_{\underline{h}} \underline{L}^{\underline{h}} \underline{I}^{\underline{h}}_h$$

and

$$\underline{L}^{2h} = \underline{I}^{2h}_{\underline{h}} \underline{L}^{\underline{h}} \underline{I}^{\underline{h}}_{2h}.$$

Noting the definitions for restriction as adjoints of the interpolation operators, we see that the variational conditions are satisfied between all grid levels, not just between h and $2h$ as certified in (1.34) and (1.35).

1.7.   Error measures.

An important but often underemphasized issue in numerical computation is the choice of error measures, which are used to assess and, in some cases, control algorithm performance. In many circumstances, this assessment can be very sensitive to the specific choices made (although this tends to be less so for multigrid than it is for other methods, as we will shortly argue). In any case, it is important that error measures be chosen that properly reflect the "real" goal of computation. Determining this goal is often subjective and heuristic, and almost always problem dependent, so there is little that can be said in any general or absolute sense. However, the importance of error measures in practice compels us to make a few observations. Although we include several relevant aspects in this informal discussion, our main objective is to make a few specific points about the nature of error measures for multilevel projection methods.

While error in numerical simulation comes from many sources, our main concerns here are the errors arising from the discretization and from the algebraic solver. More precisely, in subspace terms, let $u^* \in H_1$ be a solution of either form, (1.5) or (1.12), of the problem defined on $H = H_1$; let $u^{h^*} \in S^h \subset H$ be a solution of the discrete problem, (1.14) or (1.18); and let $u^h \in S^h$ be an approximation computed at some stage of the iteration process applied to the discrete problem. Then the error that concerns us here is the *actual error*

$$e = u^h - u^*,$$

which is the sum of the *discretization error*

$$e^* = u^{h^*} - u^*$$

and the *algebraic error*

$$e^h = u^h - u^{h^*}.$$

We first discuss the problem of assessing directly the actual error, e.

For problems posed in variational form, (1.12), a natural measure of e is derived by considering the deviation of the functional form, $R(u^h)$, from its minimum, $R(u^*)$:

$$E_R(e) \equiv R(u^h) - R(u^*).$$

Note that

$$E_R(e) = R(u^* + e) - R(u^*),$$

which shows that $E_R(e)$ does in some sense estimate the error, e. (For simplicity here and elsewhere, we restrict our discussion to *absolute* error measures, which means that we must be careful to keep the problem scale in mind in order to correctly interpret the values that are obtained. It is often better to use a *relative* error measure, which divides the absolute measure by a quantity like $R(u^*)$. This usually makes the measure independent of scale, but the appropriate relative error measure is dependent on the problem; e.g., when $R(u^*) = 0$, we must use another factor that correctly expresses the goal of computation.)

$R(u^*)$ generally cannot be computed exactly in practice, so performance tests using this measure must in most cases either estimate $R(u^*)$ numerically, or be restricted to artificial cases where $R(u^*)$ is known. Suitable estimates of $R(u^*)$ can be obtained by requiring accuracy from the numerical scheme that is much greater than the accuracy in $R(u^h)$. This generally means significantly decreasing the mesh size and dramatically increasing the number of algebraic iterations. Using this approach to assess performance amounts to comparing the approximations computed in practical tests (e.g., using realistic mesh sizes and a few iterations of the algebraic solver) to an approximation computed using many more algebraic iterations on a much finer grid. These tests often involve assessing error on a sequence of practical mesh sizes with the goal of confirming a certain order of convergence (e.g., $E_R(e) \leq ch$ for some $c < \infty$ independent of h). This approach is usually simple to implement, but it more or less begs the question by assuming beforehand that the scheme actually converges (in both the discretization and algebraic senses). This difficulty can be circumvented by estimating $R(u^*)$ using other discretization and algebraic solver schemes that are known to provide accurate, if not fast, results.

A strategy that is often safer and more useful is based on constructing a testbed of problems where the minimum is known. The ability to do this depends on the specific form of $R(u)$, however. For the first prototype problem, (1.1), this can be done by choosing a suitably smooth function u^* that is zero on the boundary of the unit square, then forming $f \equiv Lu^*$. The resulting functional $R \equiv R_L$, defined in (1.10) for this given f, then has u^* as its minimizer.

Moreover, the minimum value is just $R(u^*) = -\langle u^*, f \rangle$, which can often be determined analytically, or arbitrarily well by quadrature, for a broad class of choices of u^*. Note that the definition of e, the bilinearity of the inner product, and the fact that $Lu^* = f$ together imply, for the functional defined in (1.10), that

$$
\begin{aligned}
\ell(e) &= \ell(u^h - u^*) \\
&= R(u^h) + \langle u^*, f \rangle \\
&= R(u^h) - R(u^*) \\
&= E_R(e).
\end{aligned}
$$

The significance of this observation is that using the performance measure $E_R(e)$ is equivalent to computing $|||e|||^2$, where $||| \cdot ||| : H \to \Re$ is the *energy norm* defined by

$$
|||w||| = \ell^{1/2}(w), \quad w \in H.
$$

Note that these comments hold for problems of the form (1.1) based virtually on any linear self-adjoint operator $L : H \to H$, provided the energy norm is defined accordingly.

The second prototype problem, (1.2), is simpler because its solution is known. In fact, $u^*(x, y) \equiv \sin(\pi x) \sin(\pi y)$ is an eigenvector of $L = -\Delta$ belonging to the smallest eigenvalue, $\lambda = \pi^2$, which is the minimum value of $R = R_E$ defined by (1.11). Unfortunately, the smallest eigenvalue of operators other than the Laplacian are seldom known, so one must generally rely on somewhat less secure performance measures for these cases. Even so, confidence in the effectiveness of an algorithm can be built by observing an anticipated order of convergence of an error measure based on some strategy for estimating $R(u^*)$, especially if these observations are backed by theory.

The ultimate goal in the algorithm design process is to develop an efficient scheme that produces an actual error of acceptable size. In this process, it is important to realize that the actual error consists of discretization and algebraic errors, which is expressed by the relation

$$
e = e^* + e^h.
$$

If design begins by assessing the actual error of a practical scheme directly, then any possible deficiencies in performance cannot be easily traced to their sources. It is therefore important to be able to measure the individual error contributions from the discretization and the algebraic approximations. This would allow algorithm design to focus individually on the discretization and algebraic solver strategies.

A natural measure of the algebraic error for the variational problem, (1.12), is given by

$$
E_R^h(e^h) = R^h(u^h) - R^h(u^{h^*}).
$$

Note, as above, that

$$E_R^h(e^h) = R^h(u^{h^*} + e^h) - R^h(u^{h^*}).$$

Again we may restrict performance tests to cases where $R(u^{h^*})$ is known. This is easy to do for the linear prototype, (1.1), because we can simply select any $u^{h^*} \in S^h$ and compute $f^h = L^h u^{h^*}$ by a matrix-vector product. Here, $L^h = P_{S^h} L P_{S^h}$ (in the weak sense). This determines R^h and its minimum, $R^h(u^{h^*}) = -\langle u^{h^*}, f^h \rangle$. Analogous to the actual error case, we note that this measure of the algebraic error is related to its energy norm as follows:

$$E_R^h(e^h) = |||e^h|||^2.$$

In fact,

(1.36) $$E_R^h(e^h) = |||e^h|||_h^2,$$

where the *discrete energy norm* $||| \cdot |||_h : S^h \to \Re$ is defined by

$$|||w^h|||_h = \langle w^h, L^h w^h \rangle^{1/2}, \quad w^h \in S^h.$$

These comments so far carry over virtually to any linear self-adjoint problem of the form (1.1).

Again, the eigenvalue prototype problem, (1.2), is simpler because the smallest eigenvalue for the generalized discrete eigenproblem, (1.30), is known. In fact, a discrete eigenvector $u^{h^*} \in S^h$ can be constructed simply by *interpolating* the differential eigenvector given by $u^*(x,y) = \sin(\pi x)\sin(\pi y)$; that is, u^{h^*} is the unique element of S^h that agrees with u^* at the nodes of Ω^h. Simple trigonometric identities then show that the smallest eigenvalue is

$$\lambda^h = R^h(u^{h^*}) = \frac{36(2 - \cos(\pi h) - \cos^2(\pi h))}{h^2(4 + 4\cos(\pi h) + \cos^2(\pi h))},$$

which is an order h^2 approximation to $\lambda = \pi^2$, the smallest eigenvalue of (1.2). However, as in the actual error case, the smallest eigenvalues of other operators are seldom known. Thus, for problems of the form (1.12), we may in general be forced to use numerical estimates of $R(u^{h^*})$.

Analogous to the approach for actual error estimation, $R^h(u^{h^*})$ can be approximated by requiring more accuracy from the numerical scheme than is needed in practice, which translates in this case simply to more iterations of the algebraic solver. Again, this can be tricky because it presumes that the algebraic method converges, although other algebraic schemes that are known to converge may be used to confirm these estimates.

The discretization error, $e^* = u^{h^*} - u^*$, associated with the variational form, (1.12), may be measured by the quantity

$$E_R(e^*) \equiv R(u^{h^*}) - R(u^*).$$

To do this, we must now estimate both $R(u^{h^*})$ and $R(u^*)$, which can be done numerically by computing very accurate solutions on level h and on a much finer mesh, respectively. Once more it is important to be sure that these estimates are relatively accurate and not contaminated by weakness of the numerical scheme.

Error measures other than the energy functional may of course be used. For example, if $\| \cdot \|$ denotes some norm on H_1 (e.g., L_2, energy, or Sobolev), then $\|e\|$ may be used to measure the actual error, $\|e^*\|$ the discretization error, and $\|e^h\|$ the algebraic error. Again, this is likely to involve estimates of the exact solutions, u^* and u^{h^*}, or restriction to cases where they are known, so care must be taken with these measures.

To improve confidence in these approximations, and to enhance error estimation generally, still other error measures may be considered. For variational problems of the form (1.12), the optimal solution is often a local extremum, that is, a solution of the gradient equation

$$(1.37) \qquad\qquad \nabla R(u) = 0, \quad u \in H_1.$$

This duality between minima of the variation and zeros of the gradient can be exploited to the benefit of the numerical solution of (1.12) by introducing error measures based on (1.37). Let $\| \cdot \|$ be some norm on H_2. Then one possibility is to use the *residual norm*:

$\|\nabla R(u)\|$ measures actual error;

$\|\nabla R(u^{h^*})\|$ measures discretization error; and

$\|\nabla R^h(u^h)\|$ measures algebraic error.

These measures are appropriate when (1.37) can be taken in the strong sense. However, for the prototype problems and in many other instances, only the weak form of (1.37) holds. In such situations, we may still use the residual norm by way of the nodal vector representation: letting $\underline{r}^h = \nabla \underline{R}^h(\underline{u}^h)$ denote the residual for (1.37) and $\| \cdot \|$ the Euclidean norm on $\Re^h$, then $\|\underline{r}^h\|$ measures the algebraic error in the approximation, u^h, of the solution of (1.12). This measure has the practical significance that it is generally computable (assuming $\nabla \underline{R}^h(\underline{u}^h)$ is), so it can be used in production codes to assess and control performance.

Unless the residual norm is used with care, however, it can give misleading results. To illustrate this potential difficulty, consider the linear prototype problem posed in its variational form, (1.10). Note that the residual error norm for the nodal gradient equation is

$$\|\underline{r}^h\| = \|2(\underline{L}^h \underline{u}^h - \underline{f}^h)\|,$$

where $\underline{L}^h$ is the matrix whose entries are given by (1.28). Using the relation $\underline{L}^h \underline{u}^h - \underline{f}^h = \underline{L}^h \underline{e}^h$, then the algebraic residual error norm is given by

$$(1.38) \qquad\qquad \|r^h\| = 2\langle \underline{e}^h, (\underline{L}^h)^2 \underline{e}^h \rangle^{1/2}.$$

On the other hand, the algebraic energy error norm is given by

$$(1.39) \qquad |||\underline{e}^h|||_h = \langle \underline{e}^h, \underline{L}^h \underline{e}^h \rangle^{1/2}.$$

Now the real objective of the numerical process for solving (1.10) is presumably to achieve an acceptably small energy error. (This assumption is founded on the relation between the energy measure and energy error norm, (1.36), and the definition of the energy measure in terms of the variation.) This leads to the *ideal* convergence criterion

$$(1.40) \qquad |||\underline{e}^h|||_h < \varepsilon,$$

where $\varepsilon > 0$ is some suitably small error tolerance. (ε may correspond to any available estimate of the discretization error, $|||e^*|||$, or otherwise to a fixed reference, like machine epsilon or precision.) The energy error norm cannot be computed in most practical circumstances, so we are generally left with estimating it by the residual error norm. The point we want to make is that $\|\underline{r}^h\|$ yields a poor estimate of $|||\underline{e}^h|||_h$. To see this, note that

$$\frac{\|\underline{r}^h\|}{|||\underline{e}^h|||_h} = 2 \left(\frac{\langle \underline{v}^h, \underline{L}^h \underline{v}^h \rangle}{\langle \underline{v}^h, \underline{v}^h \rangle} \right)^{1/2},$$

where $\underline{v}^h = (\underline{L}^h)^{1/2} \underline{e}^h$, which yields the sharp estimate

$$2\gamma \le \frac{\|\underline{r}^h\|}{|||\underline{e}^h|||_h} \le 2\rho,$$

where γ and ρ are the smallest and largest eigenvalues of $(\underline{L}^h)^{1/2}$, respectively. Noting that eigenvectors for these extremal eigenvalues can be formed from the nodal values of the respective functions $\sin(\pi x)\sin(\pi y)$ and $\sin(n\pi x)\sin(\pi y)$, it is easy to see that $\gamma = \pi h$ and $\rho = 2$ (up to $O(h^2)$ relative error), hence,

$$(1.41) \qquad 2\pi h \le \frac{\|\underline{r}^h\|}{|||\underline{e}^h|||_h} \le 4 \quad \text{(up to } O(h^2) \text{ relative error).}$$

This sharp estimate shows that the residual error norm is a very weak measure of $|||\underline{e}^h|||_h$: (1.40) is satisfied *in general* only if we apply the very strict convergence criterion

$$(1.42) \qquad \|\underline{r}^h\| < 2\pi h \varepsilon \quad \text{(up to } O(h^2) \text{ relative error),}$$

even though (1.40) may already be satisfied for some error with a residual norm that is a factor of $\frac{2}{\pi h}$ larger. The implication of (1.42) is that the practical tolerance must be *much* smaller than what might first be expected.

Note that if the real objective is to achieve a small Euclidean error norm,

$$\|\underline{e}^h\| < \varepsilon,$$

then the situation is even worse. This follows from the relation

$$2\gamma^2 \leq \frac{\|\underline{r}^h\|}{\|\underline{e}^h\|} \leq 2\rho^2,$$

which leads to the yet stricter convergence criterion

$$\|\underline{r}^h\| < 2\pi^2 h^2 \varepsilon.$$

A natural mistake in assessing performance is to assume that a fairly small relative residual error signifies acceptable accuracy. This is especially hazardous in connection with relaxation methods because they quickly damp the eigencomponents of the error belonging to large eigenvalues, and the resulting high degree of smoothness means that the residuals are very small relative to the energy errors. Thus, the ratio of error norms in (1.41) tends toward the lower bound. This means that the practical convergence criterion, (1.42), is almost necessary for relaxation methods.

This stringent convergence criterion tends to be unnecessary for multigrid, partly because of its better balancing of the various eigencomponents of the error. The basic motive for multigrid is to use coarser levels to attenuate those components that cannot be adequately treated by relaxation on the finest level. Thus, after coarse-grid correction, the algebraic error tends to consist of eigencomponents belonging to the upper end of the spectrum, making the residual error norm roughly comparable to the energy error norm (in fact, even the Euclidean error norm, $\|\underline{e}^h\|$). This makes the residual error norm a reliable practical indicator of performance, provided it is used for a multigrid scheme that has already been certified to perform in this way. More importantly, the efficient and predictable behavior of such multigrid algorithms allow for a priori determination of the necessary number of iterations, which virtually eliminates reliance on dynamic convergence criteria such as (1.42). This is an often overlooked but nonetheless important advantage that multigrid has over other iterative methods. (Strategies for predicting this type of "direct" performance are discussed in Chapter 3, §3.4.)

Another type of convergence criterion involves measuring the difference of successive approximations. For example, the Euclidean norm of the difference of successive iterates produced by the algebraic solver is often used to assess and control performance. Note that this measure is closely related to the residual error norm, at least for iterative methods that use the residual to correct current approximations. This includes many relaxation methods and, in a less direct way, multigrid methods. In any case, most of the above comments made for residual error norms hold for this type of measure. Note also that it can be used to estimate discretization and actual errors by comparing approximations on different grids.

One of the major advantages of the energy error measure stems from *monotonicity* properties of the functional values. For example, we know by definition that

$$R(u^h) \geq R(u^{h^*}) \geq R(u^*),$$

so

$$E_R(e) \geq E_R(e^*).$$

It can similarly be shown that the discretization error, $E_R(e^*)$, cannot increase as the mesh is refined. Moreover, we can design the algebraic solution process so that no individual correction step can increase the error: we can always replace the generic correction step

$$u^h \leftarrow u^h + d^h$$

by the modified form

$$u^h \leftarrow u^h + sd^h,$$

where $s \in \Re$ is chosen so that

$$R^h(u^h + sd^h) \leq R(u^h).$$

We can go even further than this to achieve a sense of optimality by defining s as the minimizer of $R^h(u^h + td^h)$ over $t \in \Re$. This optimality property is actually satisfied automatically by some algebraic solvers, including the basic processes used in the multilevel projection methods developed in the next chapter. Another example of the benefits of this monotonicity property is that, while intergrid interpolation error can be a concern elsewhere, it is not in this context simply because there is none: $R^h(u^{2h}) = R^{2h}(u^{2h})$, which implies that the energy error measures of u^{2h} on levels $2h$ and h are equal. A final example is that a proper local refinement process can never impair the discretization error: a conforming local fine grid can only add elements to the discrete space, so the discretization error cannot increase.

The discussion thus far in this section has been concerned with the variational problem, (1.12). With two major exceptions, these comments apply equally well to the strong form of the equation, (1.5), using the following error measures:

$$E_K(e) = \|K(u^h)\| \quad (actual),$$
$$E_K(e^*) = \|K(u^{h^*})\| \quad (discretization), \text{ and}$$
$$E_K^h(e^h) \equiv \|K^h(u^h)\| \quad (algebraic),$$

where $\| \cdot \|$ is some norm on H_1. The exceptions are that we have gained the advantage of not having to estimate u^* or u^{h^*} and lost the advantage of monotonicity. Note that these error measures are basically residual error norms. Note also that if other performance measures like $\|e\|$ or $\|e^h\|$ are desired, then estimates of u^* or u^{h^*} may again be required. When E_K or E_K^h is used to estimate such measures, it is important to be aware of the potential weakness of this estimate, similar to the weakness in the residual error norm for the variational case. Finally, for instances where (1.5) has an equivalent variational form, this duality might be exploited to provide alternate error measures involving the variation.

Performance measures for the weak formulation, (1.9), are problematic. One possibility involves evaluating the form in terms of a basis for the space in question. Consider first the algebraic error, $e^h = u^h - u^{h^*}$, and let $\{w_i^h\}_{i=1}^n$ be a basis for T^h. Define the *residual* vector $\underline{r}^h \in \Re^n$ by its entries

$$(1.43) \qquad\qquad r_i^h = k^h(w_{(i)}^h, u^h), \quad 1 \le i \le n.$$

Using the Euclidean norm $\|\cdot\|$ on $\Re^h$, then the *algebraic residual error norm* $\|\underline{r}^h\|$ may serve as a measure of the algebraic error. In the very typical case that $k^h(v^h, u^h)$ is linear in v^h, this algebraic error measure is legitimate in the sense that it is zero only when the error is zero. This follows from observing that (1.16) is equivalent to

$$k^h\left(\sum_{i=1}^n v_i^h w_{(i)}^h, u^h\right) = 0 \quad \forall \underline{v}^h \in \Re^n,$$

and remembering that the $w_{(i)}^h$ form a basis for T^h. For the linear prototype, $k^h(v^h, u^h)$ is linear in both v^h and u^h, and since $u^h = u^{h^*} + e^h$ and $k^h(v^h, u^{h^*}) = 0$ for all $v^h \in T^h$, then

$$\|\underline{r}^h\| = \left(\sum_{i=1}^n (k^h(w_{(i)}^h, e^h))^2\right)^{1/2}.$$

Writing $e^h = \sum_{j=1}^n e_j^h w_{(j)}^h$, then

$$\|\underline{r}^h\| = \left(\sum_{i=1}^n \left(\sum_{j=1}^n k_{ij}^h e_j^h\right)^2\right)^{1/2},$$

where

$$k_{ij}^h = k^h(w_{(i)}^h, w_{(j)}^h).$$

In the case that the $w_{(i)}^h$ correspond to the hat functions defined in §1.5, the k_{ij}^h are just the entries of $\underline{L}^h$ defined in (1.28), so this residual error norm corresponds to the algebraic energy error measure defined in (1.36).

Measures for the actual or discretization errors corresponding to the weak formulation, (1.9), are much more difficult to define. We could in principle construct the residual error measure in much the same way: given $u^h \in S^h$, let $\{w_{(i)}\}_{i=1}^{\infty}$ form a basis for H_2 and define

$$r_i = k(w_{(i)}, u^h), \quad 1 \le i < \infty;$$

using the Euclidean norm $\|\cdot\|$ on $\Re^\infty$; then the *actual residual error norm* can be defined formally as $\|\underline{r}\|$. The question is now whether $\|\underline{r}\|$ is really defined, which raises the issues of what we really mean by $\Re^\infty$ and whether $\underline{r}$ is then an element of $\Re^\infty$. However, we will not pursue this further because having a well-defined actual (or discretization) error measure is more or less a theoretical question. (For practical performance assessment, we would likely approximate the actual error by replacing u^* by a solution computed on a very fine grid, and here the error measure is much like that defined in (1.43).)

CHAPTER 2

Multilevel Projection Methods

This chapter starts in §2.1 by developing the general form of the multilevel projection method, which will provide a framework that includes multigrid for uniform grids (§2.2) and the fast adaptive composite grid method (FAC) for use with local refinement (§2.3). These methods are illustrated in terms of our prototype problems in §2.4. The chapter finishes with four sections concerning practical issues and a section that summarizes the methodology.

2.1. Abstract framework: The multilevel projection method (PML).

To describe the general case, suppose S^h and S^{2h} are nontrivial finite-dimensional subspaces of H_1, and T^h and T^{2h} are nontrivial finite-dimensional subspaces of H_2. We continue to assume that these spaces are conforming, which is characterized by (1.19) in the global-grid case and by (1.20) in general. Remember that, for conventional finite element applications, these rather technical assumptions are made to ensure that the artificial boundary of the local fine grid coincides with global coarse-grid lines, and that the global coarse grid restricted to the refinement region is a subgrid of the local fine grid. In other words, the patch aligns with the global grid. Consider now the composite grid subspaces

$$(2.1) \qquad S^{\underline{h}} = S^{2h} + S^h \quad \text{and} \quad T^{\underline{h}} = T^{2h} + T^h.$$

Then the composite grid problem is written either in the equation form

$$(2.2) \qquad K^{\underline{h}}(u^{\underline{h}}) = 0, \qquad u^{\underline{h}} \in S^{\underline{h}},$$

in the variation form

$$(2.3) \qquad R^{\underline{h}}(u^{\underline{h}}) = \min_{v^{\underline{h}} \in S^{\underline{h}}} R^{\underline{h}}(v^{\underline{h}}), \qquad u^{\underline{h}} \in S^{\underline{h}},$$

or in the weak equation form

$$(2.4) \qquad k^{\underline{h}}(v^{\underline{h}}, u^{\underline{h}}) = 0, \qquad u^{\underline{h}} \in S^{\underline{h}}, \, \forall \, v^{\underline{h}} \in T^{\underline{h}}.$$

Here, $K^{\underline{h}}(u^{\underline{h}}) \equiv P_{T^{\underline{h}}} K(P_{S^{\underline{h}}} u^{\underline{h}})$, $R^{\underline{h}}(u^{\underline{h}}) \equiv R(P_{S^{\underline{h}}} u^{\underline{h}})$, and $k^{\underline{h}}(v^{\underline{h}}, u^{\underline{h}}) \equiv k(P_{T^{\underline{h}}} v^{\underline{h}}, P_{S^{\underline{h}}} u^{\underline{h}})$, $u^{\underline{h}} \in S^{\underline{h}}$, $v^{\underline{h}} \in T^{\underline{h}}$.

The two basic components of multilevel methods for solving either (2.2), (2.3), or (2.4) are *relaxation* and *coarsening*. Since we are focusing for simplicity on two-level PML algorithms, the coarsening process will be constructed in terms of an exact solver, so it is well defined, at least in principle. On the other hand, relaxation is a generic term that refers to a wide variety of approximate solvers. To be all inclusive, we would therefore have to describe relaxation in terms too general to be of much value to our development. Instead, we will restrict our use of the term relaxation to mean *block Gauss-Seidel* iteration. Actually, what we are about to develop is a generalization of block Gauss-Seidel, but this will still exclude many conventional relaxation schemes from the formulation. The reader should easily recognize that the concepts we develop do, however, allow for virtually any relaxation scheme, including other stationary linear one-step methods such as block Jacobi iteration and incomplete factorization techniques, as well as non-stationary methods such as steepest descent and multistep methods such as conjugate gradients.

To develop the generalized block Gauss-Seidel form of relaxation, let the *block* spaces S_ℓ^h, $1 \leq \ell \leq m$, constitute a *generator* for S^h, by which we mean that they are subspaces of S^h satisfying

$$(2.5) \qquad S^h = \sum_{\ell=1}^{m} S_\ell^h.$$

(Thus, any element of S^h can be written as a not necessarily unique linear combination of elements of S_ℓ^h.) Assume also that the block spaces T_ℓ^h, $1 \leq \ell \leq m$, constitute a generator for T^h. Then the general (block Gauss-Seidel) relaxation scheme, starting with an approximation, $u^h \in S^h$, is represented by the expression $u^h \leftarrow G^h(u^h)$; for the equation, (2.2), it is defined by the steps

For each $\ell = 1, 2, \cdots, m$:

Solve

$$(2.6) \qquad P_{T_\ell^h} K^h(u^h + P_{S_\ell^h} u_{(\ell)}^h) = 0, \quad u_{(\ell)}^h \in S_\ell^h$$

and set $u^h \leftarrow u^h + u_{(\ell)}^h$;

and for the variation, (2.3), it is defined by the steps

For each $\ell = 1, 2, \cdots, m$:

Solve

$$(2.7) \qquad R^h(u^h + P_{S_\ell^h} u_{(\ell)}^h) = \min_{v_{(\ell)}^h \in S_\ell^h} R^h(u^h + P_{S_\ell^h} v_{(\ell)}^h), \quad u_{(\ell)}^h \in S_\ell^h,$$

and set $u^h \leftarrow u^h + u_{(\ell)}^h$.

(We could simplify the appearance of expressions such as (2.6) and (2.7) by dropping the projection term, P_{S^h}, which would be allowed because $P_{S^h} v^h = v^h$ for every $v^h \in S^h$; however, here and elsewhere, we will preserve these projections to emphasize that these expressions represent problems confined to the relevant subspaces.)

In §2.5 we will discuss this abstract relaxation method and its interpretation as a Gauss-Seidel scheme corresponding to the block spaces S_ℓ^h. For now, we just comment that the standard point Gauss-Seidel scheme is defined by choosing an ordering, $\ell \leftrightarrow (ij)$, between the block index ℓ and the grid point indices (ij), and setting

$$(2.8) \qquad T_\ell^h = S_\ell^h = \{w_{(ij)}^h\}^\infty, \qquad \ell = 1, 2, \cdots, m \equiv n^2,$$

where $w_{(ij)}^h$ is the hat function associated with grid point P_{ij}^h and n^2 is the size of grid Ω^h.

Equations (2.6) and (2.7) define relaxation on level h. For the local refinement case, this means that the relaxation process is restricted to the local grid, Ω^h, which is what we intend.

The coarse-grid correction process is easily defined, since it involves an exact solution on the global grid, Ω^{2h}. We represent this correction by $u^h \leftarrow C^h(u^h)$; define it for the equation, (2.2), by

Solve

$$(2.9) \qquad P_{T^{2h}} K^h(u^h + P_{S^{2h}} u^{2h}) = 0, \quad u^{2h} \in S^{2h},$$

and set $u^h \leftarrow u^h + u^{2h}$;

and define it for the variation, (2.3), by

Solve

$$(2.10) \qquad R^h(u^h + P_{S^{2h}} u^{2h}) = \min_{v^{2h} \in S^{2h}} R^h(u^h + P_{S^{2h}} v^{2h}), \quad u^{2h} \in S^{2h},$$

and set $u^h \leftarrow u^h + u^{2h}$.

The abstract form of the multilevel projection method can now be described as follows. Given an initial guess $u^h \in S^h$, then one two-level cycle is represented by $u^h \leftarrow \mathrm{PML}^h(u^h)$ and defined by

Step 1 (relaxation). $u^h \leftarrow G^h(u^h)$.

Step 2 (coarse-level correction). $u^h \leftarrow C^h(u^h)$.

For the case that PML is being applied to the equation, (2.2), we refer to it as the *multilevel Petrov-Galerkin method* (GML). For the variation, (2.3), we refer to PML as the *multilevel Rayleigh-Ritz method* (RML). Their cycles are represented by the expressions $u^h \leftarrow \mathrm{GML}^h(u^h)$ and $u^h \leftarrow \mathrm{RML}^h(u^h)$, respectively.

PML provides a general framework for defining specific algorithms. In the next two respective sections, we will develop abstract PML methods for use with global grids and with local refinement. The remaining sections of this chapter are concerned with more concrete issues relevant to this general setting.

2.2.   The multigrid method (MG).

The multigrid version of PML is specified by the conventional case that the fine-grid region is global. For the conforming grids we are considering, this means that

$$S^{2h} \subset S^h = S^{\underline{h}} \quad \text{and} \quad T^{2h} \subset T^h = T^{\underline{h}},$$

which is equivalent to saying that $\Omega^h = \Omega^{\underline{h}}$. The algorithm simplifies slightly in this case as follows. One two-level cycle of MG is represented by $u^h \leftarrow MG^h(u^h)$ and defined by

Step 1. $u^h \leftarrow G^h(u^h)$.

Step 2. $u^h \leftarrow C^h(u^h)$.

2.3.   The fast adaptive composite grid method (FAC).

FAC is simply the version of PML corresponding to the case that the fine-grid region is local. Note that we thus have

$$S^{2h} \not\subset S^h \neq S^{\underline{h}} \quad \text{and} \quad T^{2h} \not\subset T^h \neq T^{\underline{h}},$$

or, equivalently,

$$\Omega^{2h} \not\subset \Omega^h \neq \Omega^{\underline{h}}.$$

The representation of one two-level cycle of FAC is $u^{\underline{h}} \leftarrow FAC^{\underline{h}}(u^{\underline{h}})$, and its steps are exactly those used to define PML.

A special *exact-solver* form of FAC will be important as a foundation for the parallel methods developed in the next section. This version of FAC is specified by using an exact solver on the local fine grid, which is characterized in $G^{\underline{h}}$ by choosing $m = 1$, $S_1^h = S^h$, and $T_1^h = T^h$. This exact solver is defined for the equation, (2.2), by

Solve

(2.11) $$P_{T^h} K^{\underline{h}}(u^{\underline{h}} + P_{S^h} u^h) = 0, \quad u^h \in S^h,$$

and set $u^{\underline{h}} \leftarrow u^{\underline{h}} + u^h$,

and for the variation, (2.3), by

Solve

(2.12) $$R^{\underline{h}}(u^{\underline{h}} + P_{S^h} u^h) = \min_{v^h \in S^h} R^{\underline{h}}(u^{\underline{h}} + P_{S^h} v^h), \quad u^h \in S^h,$$

and set $u^{\underline{h}} \leftarrow u^{\underline{h}} + u^h$.

One cycle of the exact-solver version of FAC for the equation, (2.2), thus becomes

Step 1. Solve

$$P_{T^h} K^{\underline{h}}(u^{\underline{h}} + P_{S^h} u^h) = 0, \quad u^h \in S^h,$$

and set $u^{\underline{h}} \leftarrow u^{\underline{h}} + u^h$;

Step 2. Solve

$$P_{T^{2h}} K^{\underline{h}}(u^{\underline{h}} + P_{S^{2h}} u^{2h}) = 0, \quad u^{2h} \in S^{2h},$$

and set $u^{\underline{h}} \leftarrow u^{\underline{h}} + u^{2h}$.

For the variation, (2.3), this becomes

Step 1. Solve

$$R^{\underline{h}}(u^{\underline{h}} + P_{S^h} u^h) = \min_{v^h \in S^h} R^{\underline{h}}(u^{\underline{h}} + P_{S^h} v^h), \quad u^h \in S^h,$$

and set $u^{\underline{h}} \leftarrow u^{\underline{h}} + u^h$;

Step 2. Solve

$$R^{\underline{h}}(u^{\underline{h}} + P_{S^{2h}} u^{2h}) = \min_{v^{2h} \in S^{2h}} R^h(u^{\underline{h}} + P_{S^{2h}} v^{2h}), \quad u^{2h} \in S^{2h},$$

and set $u^{\underline{h}} \leftarrow u^{\underline{h}} + u^{2h}$.

Note that this version of FAC thus amounts to alternating between exact corrections from the fine and coarse levels. As such, FAC can itself be interpreted as a block *Gauss-Seidel-like scheme* corresponding to the block subspaces S^{2h}, S^h, T^{2h}, and T^h.

2.4. Prototype problems.

Here, for concreteness, we develop specific forms of MG and FAC for treating the two prototypes, the linear elliptic PDE, (1.1), and the associated eigenvalue problem, (1.2). We first focus on the global-grid case (MG) for both prototypes, starting with the linear problem posed in its variational form, (1.12):

$$(2.13) \qquad R(u) = \min_{v \in H} R(v), \quad u \in H,$$

where $R(u) \equiv \langle \nabla u, \nabla u \rangle - 2 \langle u, f \rangle$, $u \in H = H_1$, $f \in H_2$. The discrete problem on $S^{\underline{h}} = S^{2h} + S^h$ that we are attempting to solve is

$$(2.14) \qquad R^{\underline{h}}(u^{\underline{h}}) = \min_{v^{\underline{h}} \in S^{\underline{h}}} R^{\underline{h}}(v^{\underline{h}}), \quad u^{\underline{h}} \in S^{\underline{h}},$$

where $R^{\underline{h}}(u^{\underline{h}}) = R(P_{S^{\underline{h}}} u^{\underline{h}})$, $u^{\underline{h}} \in S^{\underline{h}}$. As in §1.5 of Chapter 1, piecewise bilinear functions are used to define the subspaces for our discretization.

Since we are considering first the global-grid case, then $S^{2h} \subset S^h = S^{\underline{h}}$. The point Gauss-Seidel relaxation scheme, (2.7), which uses the one-dimensional block subspaces defined in (2.8), can be simplified by noting that the objective in each step is just to choose the optimal coefficient of the hat function, $w^h_{(ij)}$, to correct u^h. The simplified scheme is

For each $i, j = 1, 2, \cdots, n$:

Solve

(2.15) $\qquad R^h(u^h + sw^h_{(ij)}) = \min_{t \in \Re} R^h(u^h + tw^h_{(ij)}), \qquad s \in \Re,$

and set $u^h \leftarrow u^h + sw^h_{(ij)}$.

To see how s can be determined, note that the bilinearity of the inner product implies that

$$R^h(u^h + tw^h_{(ij)}) = R^h(u^h) + 2r^h_{(ij)}t + d^h_{(ij)}t^2,$$

where $r^h_{(ij)} = \langle \nabla u^h, \nabla w^h_{(ij)} \rangle - \langle w^h_{(ij)}, f \rangle$ and $d^h_{(ij)} = \langle \nabla w^h_{(ij)}, \nabla w^h_{(ij)} \rangle$. Hence, the optimal step size is given by

(2.16) $\qquad\qquad\qquad\qquad s = \dfrac{-r^h_{(ij)}}{d^h_{(ij)}}.$

It is now easy to see that this is just the usual point Gauss-Seidel method by translating the terms defining s to terms involving the nodal representation for u^h given by (1.25); this translation shows that $r^h_{(ij)}$ is just the residual of the equation $\underline{L}^h \underline{u}^h = \underline{f}^h$ at point P_{ij} $(r^h_{(ij)} = (\underline{L}^h \underline{u}^h)_{(ij)} - \underline{f}^h)$ and that $d^h_{(ij)}$ is the corresponding diagonal entry of $\underline{L}^h$ $(d^h_{(ij)} = \frac{8}{3})$.

The coarse-grid correction problem for the global-grid variational form of the equation is defined in principle by

(2.17) $\qquad R^h(u^h + P_{S^{2h}} u^{2h}) = \min_{v^{2h} \in S^{2h}} R(u^h + P_{S^{2h}} v^{2h}), \qquad u^{2h} \in S^{2h}.$

Remember that $R^h(u^h) = R^h_L(u^h) = \langle u^h, L^h u^h \rangle - 2\langle u^h, f^h \rangle$, where $L^h = P_{S^h} L P_{S^h}$ and $f^h = P_{S^h} f$. Then the bilinearity of the inner product can again be used, implying

$$
\begin{aligned}
R^h(u^h + P_{S^{2h}} v^{2h}) &= R^h(u^h) + 2\langle P_{S^{2h}} v^{2h}, L^h u^h \rangle \\
&\quad + \langle P_{S^{2h}} v^{2h}, L^h P_{S^{2h}} v^{2h} \rangle - 2\langle P_{S^{2h}} v^{2h}, f^h \rangle \\
&= R^h(u^h) + \langle v^{2h}, L^{2h} v^{2h} \rangle + 2\langle v^{2h}, P_{S^{2h}} r^h \rangle,
\end{aligned}
$$

where $r^h = L^h u^h - f^h$. Since $R^h(u^h)$ is constant with respect to the variable v^{2h}, then defining

(2.18) $\qquad R^{2h}_{u^h}(v^{2h}) = \langle v^{2h}, L^h v^{2h} \rangle + 2\langle v^{2h}, P_{S^{2h}} r^h \rangle, \qquad v^{2h} \in S^{2h},$

allows us to rewrite (2.17) as

$$R^{2h}_{u^h}(u^{2h}) = \min_{v^{2h} \in S^{2h}} R^{2h}_{u^h}(v^{2h}), \quad v^{2h} \in S^{2h}.$$

This is an important observation: $R^{2h}_{u^h}$ has the same form as R^h; in fact, $R^{2h}_{u^h}$ is the same as R^{2h} except that the source term is $-P_{S^{2h}} r^h$, not $P_{S^{2h}} f^h$. One consequence of this observation is that the coarse-grid correction equation is realizable. Note that the matrix representation for the coarse-grid correction takes the form

Solve

$$R_{\underline{u}^h}^{2h}(\underline{u}^{2h}) = \min_{\underline{v}^{2h} \in \Re^{2h}} R_{\underline{u}^h}^{2h}(\underline{v}^{2h}), \quad \underline{u}^{2h} \in \Re^{2h},$$

and set $\underline{u}^h \leftarrow \underline{u}^h + \underline{I}_{2h}^h \underline{u}^{2h}$.

Here, $R_{\underline{u}^h}^{2h}(\underline{v}^{2h}) = \langle \underline{v}^{2h}, \underline{L}^{2h}\underline{v}^{2h} \rangle + 2\langle \underline{v}^{2h}, \underline{I}_h^{2h} \underline{r}^h \rangle$, where $\underline{r}^h = \underline{L}^h \underline{u}^h - \underline{f}^h$. This is the usual variational coarsening scheme based on the variational conditions, (1.34) and (1.35).

This development shows that the relaxation steps and coarse-grid correction problem are easily implemented for the linear PDE, (1.1). Thus, each relaxation step can in principle be made exactly and inexpensively, and coarse-grid correction is realizable in the sense that the problem can easily be posed in terms of coarse-grid quantities, all of which follow from the simple form for R^h and the bilinearity of the inner product. As we now show, this ability to implement the relaxation and coarsening principles is also exhibited for the eigenvalue problem, (1.2); however, although the computation is still inexpensive, the development is somewhat more complicated.

The variational form of the discrete eigenvalue problem is just (2.14) with $R(u) \equiv \frac{\langle \nabla u, \nabla u \rangle}{\langle u, u \rangle}$, $u \in H$. (As before, we assume that the constraint $u \neq 0$ is understood.) Again, considering the global-grid case, relaxation is specified by (2.15), but now we have the more complicated expression

$$R^h(u^h + tw_{(ij)}^h) = \frac{\alpha + 2\beta t + \gamma t^2}{a + 2bt + ct^2},$$

where $\alpha = \langle u^h, L^h u^h \rangle$, $\beta = \langle u^h, L^h w_{(ij)}^h \rangle$, $\gamma = \langle w_{(ij)}^h, L^h w_{(ij)}^h \rangle$, $a = \langle u^h, M^h u^h \rangle$, $b = \langle u^h, M^h w_{(ij)}^h \rangle$, and $c = \langle w_{(ij)}^h, M^h w_{(ij)}^h \rangle$. The minimization of $R^h(u^h + tw_{(ij)}^h)$ over t generally involves inspecting the limiting cases $t = 0, \pm\infty$ and critical points, which are roots of the quadratic equation

$$(b\gamma - \beta c)t^2 + (a\gamma - \alpha c)t + (a\beta - \alpha b) = 0.$$

However, by using a full multigrid method (FMG; see Chapter 3, §3.4) to obtain an initial guess, we can usually guarantee that u^h will be close enough to the eigenvector to satisfy the liberal restriction

$$R^h(u^h) \leq R^h(w_{(ij)}^h)$$

(i.e., $\alpha/a \leq \gamma/c$). (Instead, this could be guaranteed by starting with the initial guess $u^h = w_{(pq)}^h$, where pq minimizes $R^h(w_{(ij)}^h)$ over ij.) In this case, it is easy to show that the optimal step size is the critical point

$$s = -\frac{(a\gamma - \alpha c) - \sqrt{(a\gamma - \alpha c)^2 - 4(b\gamma - \beta c)(a\beta - \alpha b)}}{2(b\gamma - \beta c)}.$$

Actually, a generally more accurate form for numerical evaluation purposes is

$$(2.19) \qquad s = -\frac{2(a\beta - \alpha b)}{(a\gamma - \alpha c) + \sqrt{(a\gamma - \alpha c)^2 - 4(b\gamma - \beta c)(a\beta - \alpha b)}}.$$

In any event, this shows that relaxation can be implemented for the eigenvalue problem, (1.2), and, although its cost is more significant than it is for relaxation on the linear equation, it is still fairly inexpensive. In fact, the quantities a, b, c, α, β, and γ used in (2.19) are calculated in much the same way as those in (2.16). For example, $a\beta - \alpha b$ is just the residual corresponding to the ijth equation for the matrix eigenvalue problem, (1.30).

To define the coarse-grid correction process for the eigenvalue problem, we again use the variational principle, (2.17). The bilinearity of the inner product and the present form of R yield

$$R^h(u^h + P_{S^{2h}} v^{2h}) = \frac{\langle u^h, L^h u^h\rangle + 2\langle u^h, L^h P_{S^{2h}} v^{2h}\rangle + \langle P_{S^{2h}} v^{2h}, L^h P_{S^{2h}} v^{2h}\rangle}{\langle u^h, M^h u^h\rangle + 2\langle u^h, M^h P_{S^{2h}} v^{2h}\rangle + \langle P_{S^{2h}} v^{2h}, M^h P_{S^{2h}} v^{2h}\rangle}.$$

Defining

$$(2.20) \qquad R_{u^h}^{2h}(v^{2h}) = \frac{\delta^h + \langle v^{2h}, p^{2h}\rangle + \langle v^{2h}, L^{2h}v^{2h}\rangle}{\epsilon^h + \langle v^{2h}, q^{2h}\rangle + \langle v^{2h}, M^{2h}v^{2h}\rangle}, \qquad v^{2h} \in S^{2h},$$

where $\delta^h = \langle u^h, L^h u^h\rangle$, $p^{2h} = 2P_{S^{2h}} L^h u^h$, $\epsilon^h = \langle u^h, M^h u^h\rangle$, and $q^{2h} = 2P_{S^{2h}} M^h u^h$, then (2.17) becomes

$$R_{u^h}^{2h}(u^{2h}) = \min_{v^{2h} \in S^{2h}} R_{u^h}^{2h}(v^{2h}), \qquad u^{2h} \in S^{2h}.$$

Just as in the linear equation case, $R_{u^h}^{2h}$ defined by (2.20) can be computed in terms of grid $2h$ quantities. In fact, its matrix form is

$$R_{\underline{u}^h}^{2h}(\underline{v}^{2h}) = \frac{\delta^h + \langle \underline{v}^{2h}, \underline{p}^{2h}\rangle + \langle \underline{v}^{2h}, \underline{L}^{2h}\underline{v}^{2h}\rangle}{\epsilon^h + \langle \underline{v}^{2h}, \underline{q}^{2h}\rangle + \langle \underline{v}^{2h}, \underline{M}^{2h}\underline{v}^{2h}\rangle},$$

where $\delta^h = \langle \underline{u}^h, \underline{L}^h\underline{u}^h\rangle$, $\underline{p}^{2h} = 2\underline{I}_h^{2h}\underline{L}^h\underline{u}^h$, $\epsilon^h = \langle \underline{u}^h, \underline{M}^h\underline{u}^h\rangle$, and $\underline{q}^{2h} = 2\underline{I}_h^{2h}\underline{M}^h\underline{u}^h$. However, this form for $R_{\underline{u}^h}^{2h}$ is not the same as the form for R^h: the additional terms involving these new quantities produce a generalization of the fine-grid form. ($R_{\underline{u}^h}^{2h}$ has the same form as R^h only when $\delta^h = \epsilon^h = 0$ and $\underline{p}^{2h} = \underline{q}^{2h} = \underline{0}$, i.e., when $\underline{u}^h = 0$.) This means that the coarse-grid problem is no longer a standard eigenvalue problem. It also means that we cannot dispense with a discussion of how to treat the grid $2h$ problem simply by appealing to recursion.

Nevertheless, the multigrid procedure for solving the problem on grid $2h$ is very similar to that for grid h. To see this, first note that

$$R_{u^h}^{2h}(u^{2h} + tw_{(ij)}^{2h}) = \frac{\alpha + 2\beta t + \gamma t^2}{a + 2bt + ct^2},$$

where $\alpha = \delta^h + \langle u^{2h}, p^{2h}\rangle + \langle u^{2h}, L^{2h}u^{2h}\rangle$, $\beta = \frac{1}{2}\langle p^{2h}, w^{2h}_{(ij)}\rangle + \langle u^{2h}, L^{2h}w^{2h}_{(ij)}\rangle$, $\gamma = \langle w^{2h}_{(ij)}, L^{2h}w^{2h}_{(ij)}\rangle$, $a = \epsilon^h + \langle u^{2h}, q^{2h}\rangle + \langle u^{2h}, M^{2h}u^{2h}\rangle$, $b = \frac{1}{2}\langle q^{2h}, w^{2h}_{(ij)}\rangle + \langle u^{2h}, M^{2h}w^{2h}_{(ij)}\rangle$, and $c = \langle w^{2h}_{(ij)}, M^{2h}w^{2h}_{(ij)}\rangle$. Thus, relaxation on the problems defined on grids $2h$ and h are identical, except that determination of the quantities used in (2.19) is a little more involved. A similar observation holds for coarsening since

$$R^{2h}_{u^h}(u^{2h} + P_{S^{4h}}\, v^{4h}) = \frac{\delta^{2h} + \langle v^{4h}, p^{4h}\rangle + \langle v^{4h}, L^{4h}v^{4h}\rangle}{\epsilon^{2h} + \langle v^{4h}, q^{4h}\rangle + \langle v^{4h}, M^{4h}v^{4h}\rangle}, \quad v^{4h} \in S^{4h},$$

where $\delta^{2h} = \delta^h + \langle u^{2h}, p^{2h}\rangle + \langle u^{2h}, L^{2h}u^{2h}\rangle$, $p^{4h} = P_{S^{4h}}(p^{2h} + 2L^{2h}u^{2h})$, $\epsilon^{2h} = \epsilon^h + \langle u^{2h}, q^{2h}\rangle + \langle u^{2h}, M^{2h}u^{2h}\rangle$, and $q^{4h} = P_{S^{4h}}(q^{2h} + 2M^{2h}u^{2h})$.

To summarize what has been discussed so far in this section, we have developed multilevel projection methods that apply specifically to our two prototypes, the linear equation, (1.1), and the eigenvalue problem, (1.2). This development was based on:

- two levels

- variational form

- global grids.

To be sure that the *two-level* description can be made practical, we showed that relaxation can be easily and efficiently implemented, and that the coarsening process is realizable and recursive. For the linear case, these properties follow naturally and lead to a classical multigrid scheme. The eigenvalue case is somewhat more complex, but no less realizable since coarse-grid correction can be written efficiently in terms of coarse-grid quantities; recursiveness comes from a slight generalization of the form of the functional (i.e., the Rayleigh quotient) that is being minimized.

We have restricted the discussion so far to the *variational form*, (1.12), i.e., RML. We could now proceed to develop these methods in the equation context, (1.5), i.e., GML. However, for the prototype problems considered here, the equation, (1.5), can be derived simply by setting the gradient of the functional in (1.12) to zero. It is therefore not hard to see that choosing the Galerkin form, GML (e.g., $T^h = S^h$), would lead us to the same method that we developed for the variational form, RML. That is, for the prototypes considered here, the projection methods developed in terms of the equation and the variation are equivalent, so it is unnecessary to develop GML for them.

Finally, RML is easily extended from the *global-grid* case to include local refinement, at least in principle. Whether the fine grid is local or global, in subspace form the method is the same; essential differences become evident only when the nodal vector spaces are considered. To clarify understanding, we will therefore complete this section with a short treatment of the nodal vector representation of FAC for the linear prototype, (1.1). For simplicity, we restrict this discussion to its exact-solver form.

The relaxation stage in subspace form is written in §2.1 as

$$R^{\underline{h}}(u^{\underline{h}} + P_{S^h}\, u^h) = \min_{v^h \in S^h} R^{\underline{h}}(u^{\underline{h}} + P_{S^h}\, v^h), \qquad u^h \in S^h.$$

With R defined by (1.10), this becomes a problem of minimizing

$$R^{\underline{h}}(u^{\underline{h}} + P_{S^h}\, v^h) = \|\nabla(u^{\underline{h}} + v^h)\|^2 - 2\langle u^{\underline{h}} + v^h, f \rangle$$

over $v^h \in S^h$. The nodal representation of this form is

$$R^{\underline{h}}(\underline{u}^{\underline{h}} + I^{\underline{h}}_h \underline{v}^h) = \langle \underline{u}^{\underline{h}} + I^{\underline{h}}_h \underline{v}^h,\, \underline{L}^{\underline{h}}(\underline{u}^{\underline{h}} + I^{\underline{h}}_h \underline{v}^h)\rangle - 2\langle \underline{u}^{\underline{h}} + I^{\underline{h}}_h \underline{v}^h,\, \underline{f}^{\underline{h}}\rangle$$
$$= R^{\underline{h}}(\underline{u}^{\underline{h}}) + R^h_{\underline{u}^{\underline{h}}}(\underline{v}^h),$$

where $R^h_{\underline{u}^{\underline{h}}}(\underline{v}^h) \equiv \langle \underline{v}^h,\, \underline{L}^h \underline{v}^h\rangle - 2\langle \underline{v}^h,\, I^h_{\underline{h}}(\underline{f}^{\underline{h}} - \underline{L}^{\underline{h}}\underline{u}^{\underline{h}})\rangle$. Thus, the FAC local-grid step is equivalent to minimizing $R^h_{\underline{u}^{\underline{h}}}(\underline{v}^h)$ over $\underline{v}^h \in \Re^h$. Note, therefore, that local-grid relaxation (in exact-solver form) is equivalent to the correction $\underline{u}^{\underline{h}} \leftarrow \underline{u}^{\underline{h}} + I^{\underline{h}}_h \underline{u}^h$, where $\underline{u}^h$ solves the local Dirichlet problem

$$\underline{L}^h \underline{u}^h = \underline{f}^h, \qquad \underline{u}^h \in \Re^h,$$

with $\underline{f}^h$ defined as the negative of the restriction of the composite grid residual, $\underline{r}^{\underline{h}} \equiv \underline{L}^{\underline{h}}\underline{u}^{\underline{h}} - \underline{f}^{\underline{h}}$, to the fine grid, Ω^h.

In a similar way, the coarse-grid correction step of FAC is equivalent to minimizing

$$R^{2h}_{\underline{u}^{\underline{h}}}(\underline{v}^{2h}) \equiv \langle \underline{v}^{2h},\, \underline{L}^{2h} \underline{v}^{2h}\rangle - 2\langle \underline{v}^{2h},\, I^{2h}_{\underline{h}}(\underline{f}^{\underline{h}} - \underline{L}^{\underline{h}}\underline{u}^{\underline{h}})\rangle$$

over $\underline{v}^{2h} \in \Re^{2h}$, which is equivalent to solving the global Dirichlet problem

$$\underline{L}^{2h}\, \underline{u}^{2h} = \underline{f}^{2h}, \qquad \underline{u}^{2h} \in \Re^{2h},$$

with $\underline{f}^{2h} = I^{2h}_{\underline{h}}(\underline{f}^{\underline{h}} - \underline{L}^{\underline{h}}\underline{u}^{\underline{h}})$.

2.5. Relaxation.

We have thus far discussed two examples of relaxation, point Gauss-Seidel and exact solvers. These are the extreme cases of the general relaxation process defined (for the Rayleigh-Ritz or Galerkin cases) according to the block decomposition of S^h given by (2.5). Point Gauss-Seidel is defined by choosing $m = n^2$ and $S^h_\ell = \{w^h_{(ij)}\}^\infty$, $\ell = 1, 2, \cdots, m$, for some ordering $\ell \leftrightarrow (ij)$, and exact solvers are characterized by choosing $m = 1$ and $S^h_1 = S^h$. In some applications, intermediate forms of relaxation might be more appropriate. For example, if we modify the linear prototype so that it exhibits anisotropic behavior, then line relaxation might be needed: for the linear PDE

$$-(\varepsilon u_{xx} + u_{yy}) = f,$$

with $0 < \varepsilon \ll 1$, we would use y-line relaxation, which is characterized by choosing $m = n$ and $S_\ell^h = \{w_{(\ell j)}^h : 1 \leq j \leq n\}^\infty$, $\ell = 1, 2, \cdots, n$. Other types of relaxation (e.g., distributive, collective, and box; cf. [Brandt 1984; §3.4]) can also be described by proper choice of these block subspaces.

For illustration, and because it is an important basic relaxation scheme in the multilevel context, we briefly show here how the point *Kaczmarz* method can be described in this framework. This iterative scheme can be applied to a general matrix equation

$$\underline{A}\underline{x} = \underline{b}, \qquad \underline{x} \in \Re^n,$$

where $\underline{A}$ is $m \times n$ and $\underline{b} \in \Re^m$, with $\underline{b}$ assumed to be in the range of $\underline{A}$ (so that a solution exists). Kaczmarz applied to this equation is written as

For each $\ell = 1, 2, \cdots, n$:

Determine s so that

$$\langle \underline{w}_{(\ell)}, \, \underline{A}(\underline{x} + s\underline{A}^t\underline{w}_{(\ell)}) - \underline{b} \rangle = 0$$

and set $\underline{x} \leftarrow \underline{x} + s\underline{A}^t\underline{w}_{(\ell)}$.

Here, $\underline{w}_{(\ell)}$ is the ℓth coordinate vector in $\Re^m$, which has one in the ℓth entry and zero elsewhere. Note that $\underline{A}^t\underline{w}_{(\ell)}$ is the ℓth row of $\underline{A}$. Now we can generalize this scheme to apply to the possibly nonlinear system, (2.2), by defining the Fréchet derivative $J^{\underline{h}} = (K^{\underline{h}})'(v^{\underline{h}})$ for some fixed function $v^{\underline{h}} \in S^{\underline{h}}$, yielding

For each $\ell = 1, 2, \cdots, m$:

Determine s so that

$$\langle w_{(\ell)}^h, \, K^{\underline{h}}(u^{\underline{h}} + s(J^{\underline{h}})^* w_{(\ell)}^h) \rangle = 0$$

and set $u^{\underline{h}} \leftarrow u^{\underline{h}} + s(J^{\underline{h}})^* w_{(\ell)}^h$.

Here, $(J^{\underline{h}})^*$ is the adjoint of $J^{\underline{h}}$, $w_{(\ell)}^h$ is the ℓth canonical basis vector in $T^{\underline{h}}$, and m is the dimension of $T^{\underline{h}}$. Note that Kaczmarz now seems to fit into the abstract framework of our general relaxation scheme by choosing $S_\ell^{\underline{h}} = \{(J^{\underline{h}})^* w_{(\ell)}^h\}^\infty$ and $T_\ell^{\underline{h}} = \{w_{(\ell)}^h\}^\infty$, $1 \leq \ell \leq m$.

Actually, there is a subtle *technical* difficulty here because the $S_\ell^{\underline{h}}$ do not generally span $S^{\underline{h}}$. If $K(u)$ is linear, a solution must lie in the range of $(J^{\underline{h}})^*$, since this is just the orthogonal complement of the null space of $J^{\underline{h}}$. (This follows from the fundamental theorem of linear algebra: the null space of a finite-dimensional linear operator is equal to the orthogonal complement of the range of its transpose.) Since the S_ℓ^h span this range, we could simply take $S^h \equiv S_1^h + S_2^h + \cdots + S_m^h$, which only has the beneficial effect of eliminating the nontrivial null space of K^h. However, the possible nonlinearity of K prevents us from doing this in general. Another subtlety here is that the matrix J^h would likely change in practice, chosen in terms of the vector $v^h = u^h$ at various stages of the relaxation process. Thus, Kaczmarz does not strictly fit into our formalism, although it certainly is compatible with projection methods, as §4.6 in Chapter 4 will illustrate.

No general prescription will be given here for guiding the proper choice of relaxation. In fact, this is not now possible, and it may never be, since the effectiveness of a particular relaxation scheme is very dependent on the general physical character of the underlying problem. Nevertheless, by concentrating on a given application, it is usually possible to determine a relaxation process of the type considered here that works effectively at eliminating error components not well approximated by the coarser subspaces. This should be the goal of relaxation. We will describe attempts to achieve this goal for the applications developed in Chapter 4.

The relaxation scheme in its general form is more or less a guiding principle. For the extreme case of an exact solver, we do not mean to suggest that direct matrix solvers are required in practice. In fact, we really have in mind nothing more costly than simple multigrid V-cycles, which tend to be strongly convergent iterative methods, but rarely direct solvers. Similarly, for practical use of the intermediate forms of relaxation, we almost never mean exact solvers on the subspaces; approximate line solvers are generally adequate, for example. Even for the simplest case of point relaxation, exact determination of the step size is almost never necessary in practice, which is fortunate because nonlinearity usually makes exact solution impossible. In fact, one step of a Newton or Newton-like process is often enough to determine a step size adequate for smoothing purposes. No doubt more complicated functionals require greater care to avoid wandering too far from the proper direction and succumbing to other pitfalls that may exist. But these difficulties are usually small compared to the underlying problem, particularly for point relaxation where the problem of determining the correction is one-dimensional.

2.6.   Coarse-level realizability and recursiveness.

We saw in the previous section how the general relaxation step of PML is more of a guiding principle than an algorithm: the process that it defines can be implemented exactly only in special cases. This is even more true of the coarsening step: not only is an exact solve on level $2h$ usually impractical, but just determining how the correction process can be expressed exactly in terms only of $2h$ quantities is often a difficult if not impossible task. Exact solution on level $2h$ is almost never used; in fact, it is usually enough to perform a few relaxations, then appeal once (V-cycles) or twice (W-cycles) to level $4h$ for corrections (which are themselves "solved" by relaxation and grid $8h$ correction, and so on). However, the success of this approach depends on our ability to write the correction as a level $2h$ problem (*coarse-level realizability*) that has the same form as the level h problem (*recursiveness*). This can be done for both prototype problems, as shown in §2.4, although obtaining recursiveness for the eigenvalue problem required a slight generalization of the fine-level form.

Coarse-level realizability means that the correction from the coarse level can be computed by a process that involves a reasonable amount of coarse-level calculations. Considering the general equation, (1.5), and its grid h discretization, (1.14), this means that we must be able to write $P_{T^{2h}} K^h(u^h + P_{S^{2h}} u^{2h})$ in terms of level $2h$ quantities. This is almost analogous to realizability of the discretization, which asserts that $P_{T^h} K(P_{S^h} u^h)$ can be written in terms of level h quantities. The exception is the presence of u^h. In fact, it was just this additional term that required generalizing the form of the fine-level eigenvalue problem. The point here is that both coarse-level realizability and recursiveness would be achieved if we could realize discretization of the more general form $P_{T^h} K(u + P_{S^h} u^h)$, where $u \in H$ is given.

On the other hand, exact coarse-level realization is not necessary, provided there is a way to approximate the correction problem well enough to ensure reasonable representation of "smooth" error components (or, more precisely, error components not adequately reduced by a few relaxation sweeps). This can often be done for a given application by inspecting the form of the problem and perhaps using a little physical insight. To see how this might be done more abstractly, consider the correction equation

$$(2.21) \qquad P_{T^{2h}} K^h(u^h + P_{S^{2h}} u^{2h}) = 0.$$

Assume that the discretization process is realizable (i.e., $K^h(u^h)$ and $K^{2h}(u^{2h}) \equiv P_{T^{2h}} K(P_{S^{2h}} u^{2h})$ are representable on their respective spaces, S^h and S^{2h}). Now to approximate this correction equation on grid $2h$, we could simply replace u^h in (2.21) by $P_{S^{2h}} u^h$, yielding

$$P_{T^{2h}} K^h(P_{S^{2h}}(u^h + u^{2h})) = 0,$$

or

$$(2.22) \qquad K^{2h}((P_{S^{2h}} u^h) + u^{2h}) = 0.$$

Unfortunately, besides being generally inconsistent with (1.14) ($u^h = u^{h^*}$ does not imply that $u^{2h} = 0$), (2.22) may poorly represent "smooth" error components. Consider the linear case, $K = K_L$, where (2.22) becomes

$$L^{2h} u^{2h} = f^{2h} - L^{2h}(P_{S^{2h}} u^h)$$

or

$$L^{2h} u^{2h} = P_{T^{2h}}(f^h - L^h(P_{S^{2h}} u^h)).$$

Such a form would generally not produce a good correction. This can be seen by observing that $f^h - L^h(P_{S^{2h}} u^h)$ differs from the fine-grid residual by the term $L^h(u^h - P_{S^{2h}}\, u^h)$, which has the potential effect of introducing spurious oscillatory components that contaminate the smooth ones being transferred via the residual to grid $2h$. The key to avoiding this trouble is to rewrite (2.21) so that the terms that are to be approximated using $2h$ projections are smooth (assuming we have used enough relaxation sweeps). One possible form is

$$(2.23) \qquad P_{T^{2h}}\, K^h(u^h + P_{S^{2h}}\, u^{2h}) - P_{T^{2h}}\, K^h(u^h) = -P_{T^{2h}}\, K^h(u^h).$$

Replacing u^h on the left-hand side of (2.23) by $P_{S^{2h}}\, u^h$ then yields

$$P_{T^{2h}}\, K^h(P_{S^{2h}}(u^h + u^{2h})) - P_{T^{2h}}\, K^h(P_{S^{2h}}\, u^h) = -P_{T^{2h}}\, K^h(u^h),$$

or

$$K^{2h}((P_{S^{2h}}\, u^h) + u^{2h}) - K^{2h}(P_{S^{2h}}\, u^h) = -P_{T^{2h}}\, K^h(u^h).$$

Writing $\tau_h^{2h} \equiv K^{2h}(P_{S^{2h}}\, u^h) - P_{T^{2h}}\, K^h(u^h)$, this becomes

$$K^{2h}((P_{S^{2h}}\, u^h) + u^{2h}) = \tau_h^{2h}.$$

Thus, a grid $2h$ correction scheme based on this approximation to the PML projection is given by

Solve

$$K^{2h}(v^{2h}) = \tau_h^{2h}$$

and set $u^h \leftarrow u^h + (v^{2h} - P_{S^{2h}}\, u^h)$.

This is in fact the widely used *full approximation scheme* (FAS; cf. [Brandt 1984, Chapter 8]). In this way, FAS can be viewed as an approximation to the PML coarse-level correction process, which can be used whenever PML cannot be exactly or easily realized.

2.7.   Parallelization: Asynchronous FAC (AFAC).

Conventional multilevel methods, including multilevel projection methods, seem to have one major limitation in a parallel computing environment: coarse-level correction must wait for relaxation to be completed, and conversely. This can be a concern for applications that demand many levels of discretization and many processors because this implies processor idle time, which is perceived as inefficient use of the computer. In the global-grid case (MG), the potential difficulty is that processors that may be assigned to coarser levels are useless when the finer levels are being processed. However, for most practical applications, there are many more points than processors; thus, computation on the finer levels, which have most of the grid points, is dominant and processor idleness has negligible impact. (See [Briggs, Hart, McCormick, and Quinlan 1988] for an analysis.) Even in the very extreme case of one point per processor, the effect of processor idleness is to increase computation time by a factor equal roughly to the number of levels which, for this global grid case, is on the order of the logarithm of the number of fine-grid points. This is a fairly modest penalty for such extreme circumstances, and in any case it may be unavoidable (i.e., multigrid may be optimal in parallel computation; cf. [Chan and Tuminaro 1987]).

Unfortunately, processor idleness is much less benign in the local refinement case. The difference here that induces this trouble is that, in practice, the finer levels tend to have relatively few grid points. Thus, they no longer dominate computational demands, yet they can significantly increase processing time. Said differently, the underlying motive for local refinement disappears when there are many processors because the need to schedule the levels sequentially makes the cost of treating local and global grids essentially the same. In the extreme case of one point per processor and very many levels, the number of these levels can be on the same order as the number of points, so the increase in computation time due to processor idleness can be catastrophic. Fortunately, a simple and inexpensive modification to FAC creates independence between the levels so that they may be processed simultaneously, with no essential loss in convergence rate.

To see how this is done, it is important first to determine what prevents the FAC levels from being processed simultaneously. Focusing for simplicity on the two-level exact-solver version of FAC for the variation, (2.3), suppose that we attempt to compute the fine-level and coarse-level corrections simultaneously. Remembering the block Gauss-Seidel interpretations of FAC, this leads to a block *Jacobi-like* algorithm of the form

Solve both

$$(2.24) \qquad R^{\underline{h}}(u^{\underline{h}} + P_{S^h} u^h) = \min_{v^h \in S^h} R^{\underline{h}}(u^{\underline{h}} + P_{S^h} v^h), \quad u^h \in S^h,$$

and

$$(2.25) \qquad R^{\underline{h}}(u^{\underline{h}} + P_{S^{2h}} u^{2h}) = \min_{v^{2h} \in S^{2h}} R^{\underline{h}}(u^{\underline{h}} + P_{S^{2h}} v^{2h}), \quad u^{2h} \in S^{2h},$$

then set $u^{\underline{h}} \leftarrow u^{\underline{h}} + u^h + u^{2h}$.

To determine the behavior of this scheme, recall that the defining expression for $S^{\underline{h}}$ in (2.1) is not a direct sum because $S^{2h,h} \equiv S^{2h} \cap S^h \neq \emptyset$. Let $e^{\underline{h}}$ be the algebraic error in the initial approximation, $u^{\underline{h}}$, to the exact solution, $u^{\underline{h}^*}$, of (2.3) so that

$$(2.26) \qquad u^{\underline{h}} = u^{\underline{h}^*} + e^{\underline{h}}.$$

Suppose that $e^{\underline{h}} \in S^{2h,h}$. Then $e^{\underline{h}} \in S^h$, so a solution of (2.24) is just $u^h = -e^{\underline{h}}$. Similarly, since $e^{\underline{h}} \in S^{2h}$, then a solution of (2.25) is $u^{2h} = -e^{\underline{h}}$. Hence, by (2.26), $u^{\underline{h}} + u^h + u^{2h} = u^{\underline{h}^*} - e^{\underline{h}}$. This means that *one cycle of the Jacobi-like FAC scheme changes only the sign of the approximation error in $S^{2h,h}$*. This reasoning shows that, in general, error components in $S^{2h,h}$ cannot be reduced by this approach. (Damping factors could be introduced, but to be effective they would have to be inversely proportional to the number of refinement levels, and overall convergence would then deteriorate by this factor.)

The culprit here is that the *local coarse-level* error components (i.e., the components in $S^{2h,h}$) are computed twice, once on Ω^{2h} and once on Ω^h. Correction by both levels therefore overshoots the solution. A simple cure that eliminates this redundancy is to compute these special error components separately, then subtract them out of the resulting correction. This leads us to the special modified Jacobi-like version of FAC, which we call AFAC; one two-level cycle of AFAC is represented by the expression $u^h \leftarrow \text{AFAC}^h(u^h)$ and defined by

Solve the three problems

$$R^h(u^h + P_{S^h}\, u^h) = \min_{v^h \in S^h}\ R^h(u^h + P_{S^h}\, v^h), \quad u^h \in S^h,$$

$$R^h(u^h + P_{S^{2h,h}}\, u^{2h,h}) = \min_{v^{2h,h} \in S^{2h,h}}\ R^h(u^h + P_{S^{2h,h}}\, v^{2h,h}),$$
$$u^{2h,h} \in S^{2h,h},$$

and

$$R^h(u^h + P_{S^{2h}}\, u^{2h}) = \min_{v^{2h} \in S^{2h}}\ R^h(u^h + P_{S^{2h}}\, v^{2h}), \quad u^{2h} \in S^{2h},$$

then set $u^h \leftarrow u^h + u^h - u^{2h,h} + u^{2h}$.

The scheme for solving the equation, (2.2), is defined analogously by

Solve the three problems

$$P_{T^h}\, K^h(u^h + P_{S^h}\, u^h) = 0, \quad u^h \in S^h,$$

$$P_{T^{2h,h}}\, K^h(u^h + P_{S^{2h,h}}\, u^{2h,h}) = 0, \quad u^{2h,h} \in S^{2h,h},$$

and

$$P_{T^{2h}}\, K^h(u^h + P_{S^{2h}}\, u^{2h}) = 0, \quad u^{2h} \in S^{2h},$$

then set $u^h \leftarrow u^h + u^h - u^{2h,h} + u^{2h}$.

Here we use the notation $T^{2h,h} = T^{2h} \cap T^h$.

Note that the different levels in AFAC may be processed independently of each other. This is true whether there are just two basic levels ($2h$ and h), or many. In fact, the convergence behavior of AFAC is qualitatively the same as that of FAC; the primary difference is that AFAC error reduction factors are typically the square root of FACs (which is, by no coincidence, analogous to the relationship between Jacobi and Gauss-Seidel point relaxation methods for M-matrices). Since their computational demands are comparable ($S^{2h,h}$ processing costs are negligible compared to S^h costs), AFAC takes typically twice as many operations to achieve a given convergence criterion; but its full parallelizability makes it the clear winner in very large scale parallel computer applications. (See [Lemke and Quinlan 1991] for recent computation results on this issue.)

For simplicity, because AFAC does not fit into the general framework of the PML methods, we will devote very little further discussion to it. Instead, the development will focus on the sequentially structured PML methods. Nevertheless, the reader interested in parallel computation should easily recognize the applicability of AFAC in what follows.

2.8. Other practical matters.

We have avoided discussion of several practical issues relevant to the abstract framework developed in this chapter. A more thorough development of multilevel methods should include, as a start: the various cycling strategies (e.g., V-cycles, W-cycles, and μ-cycles); *full multigrid* (FMG), or *nested iteration*, methods for obtaining optimal complexity; other versions of FAC and AFAC and the related issue of proper treatment of the composite grid interface; and the various methods for assessing efficiency and accuracy, especially a priori theoretical and practical methods of analyzing performance. These and other practical matters are treated in detail in the literature (cf. [Brandt 1984], [McCormick 1989], and [Stüben and Trottenberg 1982]), and since they are not really special to the methods developed here, we will say very little more about them.

2.9. Summary.

Here we list the essential relations for the four basic ingredients of the multilevel projection methodology. This is done for both the equation (remembering that this can be taken in the weak sense) and the variation. We present these relations in their simplest forms, dropping unnecessary projections and superscripts. (Note, for example, that $P_{T_\ell^h} K(u^h + u^h_{(\ell)}) = P_{T_\ell^h} K^h(u^h + P_{S_\ell^h} u^h_{(\ell)})$.)

equation

problem $(K : H_1 \to H_2)$

$$K(u) = 0, \quad u \in H_1.$$

discretization $(S^h \subset H_1, T^h \subset H_2)$

$$P_{T^h} K(u^h) = 0, \quad u^h \in S^h.$$

relaxation $\left(\sum_{\ell=1}^m S_\ell^h = S^h, \sum_{\ell=1}^m T_\ell^h = T^h \right)$

$$P_{T_\ell^h} K(u^h + u^h_{(\ell)}) = 0, \quad u^h_{(\ell)} \in S_\ell^h.$$

coarsening $(S^{2h} \subset S^h, T^{2h} \subset T^h)$

$$P_{T^{2h}} K(u^h + u^{2h}) = 0, \quad u^{2h} \in S^{2h}.$$

variation

problem $(R : H \to \Re)$

$$R(u) = \min_{v \in H} R(v), \quad u \in H.$$

discretization $(S^h \subset H)$

$$R(u^h) = \min_{v^h \in S^h} R(v^h), \quad u^h \in S^h.$$

relaxation $\left(\displaystyle\sum_{\ell=1}^{m} S_\ell^h = S^h \right)$

$$R(u^{\underline{h}} + u_{(\ell)}^h) = \min_{v_{(\ell)}^h \in S_\ell^h} R(u^{\underline{h}} + v_{(\ell)}^h), \quad u_{(\ell)}^h \in S_\ell^h.$$

coarsening $(S^{2h} \subset S^h)$

$$R(u^{\underline{h}} + u^{2h}) = \min_{v^{2h} \in S^{2h}} R(u^{\underline{h}} + v^{2h}), \quad u^{2h} \in S^{2h}.$$

CHAPTER 3

Unigrid

The task of developing computer code for multilevel algorithms can be very imposing. This process involves: determining and manipulating data structures; controlling cycling strategies; initializing arrays; treating boundary conditions; scaling matrices, interlevel transfers, and norms; and assessing and controlling performance. These and other aspects of implementation mandate a full understanding of the relevant concepts and processes, and they mask many pitfalls to try even the most experienced scientist. These difficulties are greatly compounded when the objective is to test the performance of several versions of the basic algorithm or of one version over a wide variety of applications. For example, starting with a basic multigrid code that was designed for use with uniform global grids, it can take many days, or more likely several weeks, of effort to test how this scheme would work in local refinement mode. We will show in this chapter that a special computational tool exists that can often reduce the time of these design tasks to just a few minutes.

The main cause of these implementation difficulties is the need to create and manipulate data structures that support computational processes on coarser levels. The real need for the coarse-level arrays is for storing intermediate results. Therefore, an easy mechanism for eliminating the major troubles of implementation is to eliminate the need for coarse-level arrays by arranging the computation so that every coarse-level calculation is immediately reflected in the fine-level arrays. This is the basic idea behind *unigrid* (UG; cf. [McCormick and Ruge 1983]). (We will continue to use the term unigrid for historical reasons, even though *unilevel* would be more consistent with our conventions.)

To clarify this basic point, remember that one of the main issues relevant to the efficiency of the PML coarsening process is the ability to translate the abstract correction problem into one that involves only coarse-level quantities. The ability to do this is what we have been calling coarse-level realizability. The objective of this translation is to relegate coarse-level processing to arrays that are much shorter than the fine-level arrays, making the cost relatively insignificant. Implementation of the coarsening process, as it is originally posed, directly on the fine level, would be considerably more expensive. However, the immediate goal of multilevel algorithm design and feasibility studies is not really

49

to produce efficient code; it is usually enough just to know that the algorithms being developed could be coded later in a fully efficient way. In other words, it is important to have a coding strategy that allows easy implementation and modification of multilevel algorithms, almost without regard to computational cost, provided the results that these codes produce accurately predict the results of the algorithms implemented in fully efficient form.

Unigrid offers just such a capability. It will be developed in this chapter as a natural consequence of applying coarse-level relaxation directly to the abstract fine-level correction principle treated in the previous chapter.

3.1.   Basic unigrid scheme.

For simplicity, consider first the global-grid discretization of the variation, (1.12), which has the form

$$(3.1) \qquad R^h(u^h) = \min_{v^h \in S^h} R^h(v^h), \quad u^h \in S^h.$$

We assume throughout most of this chapter that point Gauss-Seidel is the smoother used on all levels of the RML process. One step of this relaxation scheme on S^h can be written in *directional iteration* form as follows:

Solve

$$(3.2) \qquad R^h(u^h + sd) = \min_{t \in \Re} R(u^h + td), \quad s \in \Re,$$

and set $u^h \leftarrow u^h + sd$.

Here, d is an element of a given basis, $\{d^h_{(\ell)} : 1 \le \ell \le m^h\}$, for S^h. For the prototype problems, $d^h_{(\ell)} = w^h_{(ij)}$ using some ordering $\ell \leftrightarrow (ij)$. For convenience, we write this single step, (3.2), using a given direction $d \in S^h$ as

$$u^h \leftarrow D^h(u^h; d).$$

Obviously, relaxation consists solely of grid h computations. The objective now is to arrange things so that this is also true of all steps in the coarsening process.

The coarse-grid correction step of RML for (3.1) is given abstractly as

Solve

$$(3.3) \qquad R^h(u^h + u^{2h}) = \min_{v^{2h} \in S^{2h}} R^h(u^h + v^{2h}), \quad u^{2h} \in S^{2h},$$

and set $u^h \leftarrow u^h + u^{2h}$.

(Here we drop the use of $P_{S^{2h}}$ because we no longer wish to stress that the correction equation defined in (3.3) can be realized on grid $2h$; this omission is allowed because $P_{S^{2h}} v^{2h} = v^{2h}$ for every $v^{2h} \in S^{2h}$.) A typical step of a practical solution process on grid $2h$ would involve updating the given approximation, $u^{2h} \in S^{2h}$, by choosing d as an element of a given basis, $\{d^{2h}_{(\ell)} : 1 \le \ell \le m^{2h}\}$, of S^{2h} and performing the following:

Solve

$$R^h(u^h + u^{2h} + sd) = \min_{t \in \Re} R^h(u^h + u^{2h} + td), \quad s \in \Re,$$

and set $u^{2h} \leftarrow u^{2h} + sd$.

Here we have resisted the temptation to write this step in terms of the form $R^h_{u^h}(u^{2h}) \equiv R^h(u^h + u^{2h})$, because now it should be clear that there is no real need to preserve u^{2h}: instead of storing u^{2h} separately, any time u^{2h} is to be changed, we simply add the correction directly to u^h. This reasoning shows that using just one relaxation sweep on grid $2h$ for solving (3.3), starting with $u^{2h} \leftarrow 0$, is equivalent to the following:

For $\ell = 1, 2, \cdots, m^{2h}$:

$$u^h \leftarrow D^h(u^h; d^{2h}_{(\ell)}).$$

In fact, a little reflection shows that *all* coarse-grid relaxations are equivalent to directional relaxations on u^h. For example, suppose we have $p \geq 1$ grids of mesh sizes h_ν, $1 \leq \nu \leq p$, where $h_\nu = 2h_{\nu+1}$, $1 \leq \nu \leq p-1$. Write $h = h_p$. Suppose $\{d^{h_\nu}_{(\ell)} : 1 \leq \ell \leq m^{h_\nu}\}$ is a given basis for S^{h_ν}, $1 \leq \nu \leq p$. Then one $(0,1)$ *V-cycle* (which performs no relaxations before coarsening, one on each level afterwards) is equivalent to the following fine-grid process:

For $\nu = 1, 2, \cdots, p$:

For $\ell = 1, 2, \cdots, m^{h_\nu}$:

(3.4) $$u^h \leftarrow D^h(u^h; d^{h_\nu}_{(\ell)}).$$

Equation (3.4) is the abstract form of unigrid, which we developed for minimizing the variation, (1.12), on global grids. The form for solving the equation, (1.5), is the same except for the definition of D^h. Consider the discrete equation obtained on a global grid by the Galerkin form of GML (i.e., $T^h = S^h$),

(3.5) $$K^h(u^h) = 0, \quad u^h \in S^h.$$

Let $d \in S^h$. Then the general directional iteration scheme applied to (3.5) is again represented by $u^h \leftarrow D^h(u^h; d)$, but now defined by

Solve

$$\langle d, K^h(u^h + sd) \rangle = 0, \quad s \in \Re,$$

and set $u^h \leftarrow u^h + sd$.

If (3.5) represents the Petrov-Galerkin form of GML, then the directional iteration scheme depends on $d_S \in S^h$ and $d_T \in T^h$ and is defined by

Solve

$$\langle d_T, K^h(u^h + sd_S) \rangle = 0,$$

and set $u^h \leftarrow u^h + sd_S$.

This we write as $u^h \leftarrow D^h(u^h; d_S, d_T)$. Of course, in this case, the unigrid scheme requires bases for the T^{h_ν} as well as the S^{h_ν}, and (3.4) must be modified accordingly.

3.2. Multigrid simulation.

Implementation of (3.4) is relatively simple. The main task is to develop a general directional iteration routine, then apply this routine to the vector representations of the $d_{(\ell)}^{h_\nu}$ on grid h:

$$\underline{d} = \underline{I}_{h_\nu}^h \, \underline{d}_{(\ell)}^{h_\nu}, \quad 1 \le \ell \le m^{h_\nu}, \, 1 \le \nu \le p,$$

where $\underline{I}_{h_\nu}^h : \Re^{h_\nu} \to \Re^h$ is the interpolation operator that maps from the space of nodal vectors on grid h_ν to those on grid h. For example, as with the two prototype problems, if S^{h_ν} consists of piecewise bilinear functions, then $\underline{I}_{h_\nu}^h$ is the usual linear interpolation operator. See Figure 3.1 for a depiction of sample directions on three grid levels using bilinear interpolation in one dimension for the standard coordinate basis functions. The directions in two or more dimensions are formed as tensor products of the one-dimensional basis functions.

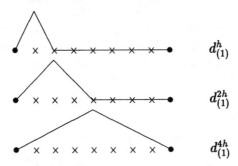

$$d_{(1)}^h$$

$$d_{(1)}^{2h}$$

$$d_{(1)}^{4h}$$

FIG. 3.1. *Sample one-dimensional unigrid directions.*

The appendices contain listings of routines for solving Poisson's equation on the unit square in $\Re^2$, which is the first prototype, (1.1). The routines in Appendix A assume a uniform global grid and discretization by continuous piecewise bilinear functions with respect to the grid cells. This yields the usual nine-point stencil given in (1.28). The entry routine in Appendix A is *unigrid*, with the form

subroutine unigrid $(n, u, f, r, d, ad, ncyc, nlvl)$.

Here:

n is the number of interior grid lines in both the x and y directions;

$nlvl$ is the number of unigrid "levels" $[2 * * (nlvl - 1)$ must divide $n + 1$; e.g., $n = 31$ and $nlvl = 6]$;

$ncyc$ is the number of $(0,1)$ V-cycles to be performed;

u is the array containing the initial approximation on input, and the final result on output;

f is the source term array; and

r, d, ad are work arrays.

All arrays are to be dimensioned $(0 : n + 1; 0 : n + 1)$. The unigrid routine can thus be called by a driver that defines n, $nlvl$, $ncyc$, u, and f and simply dimensions r, d, and ad.

The subroutines in Appendix A have the following roles:

$unigrid(n, u, f, r, d, ad, ncyc, ulvl)$. This is the primary routine that implements the cycling process and performs the directional iteration scheme, (3.4).

$amult(n, u, av)$. This routine applies the fine-grid stencil defined in (1.28) to the input vector u and places the result in array au. Other stencils, and other PDEs, can be treated simply by changing the assignment statement in $amult$.

$direction(n, d, i, j, m)$. Here, the hat function centered at point (i, j) of the fine grid, with support at points $(i + k, j + \ell)$, $|k|, |\ell| \le m - 1$, is computed and stored in d. (m is the mesh factor, which is the mesh size for the level associated with d divided by the finest mesh size.)

$dot(n, x, y, xdoty)$. This is a BLAS-like routine that forms the dot product of the two-dimensional arrays x and y and stores it in the scalar $xdoty$.

$error(h, u, f, r, kcyc)$. This routine prints out the Euclidean norm of the residual error for cycle number $kcyc$.

$xpcy(n, x, y, c)$. This is a BLAS-like routine that takes the two-dimensional arrays x and y and the scalar c and forms $x + c * y$, storing the result in x.

$zero(n, x)$. This zeros out the two-dimensional array x.

The directions $d_{(\ell)}^{h_\nu}$, which are computed in the subroutine $direction$, are not assumed to have any special zero structure. This greatly simplifies the $unigrid$ routine, but it is highly inefficient, especially on the finest level where all but one of the direction components are zero. Efficiency is not a primary concern for many unigrid purposes, so this type of unigrid is likely to be useful, at least in the early stages of algorithm design. However, multilevel performance tests ultimately involve studying the dependence of convergence rates on the mesh size, which requires at least moderate-scale discretizations. The larger grids in these tests are likely to expose the inefficiency of this version of unigrid to the extent that its cost can become prohibitive.

For such larger-scale uses, it would be advisable to develop a more efficient version of unigrid that accounts for the zero structure of the direction vectors. Appendix B contains a listing of a collection of routines of this type, applied again to the linear prototype, (1.1). The increased code complexity of these routines is the result of manipulating indices to avoid computation involving zeros of the directions. These directions are generated in a template fashion: on a given level, since all hat functions are the same up to coordinate translation, only one reference hat function is generated; all others are formed by a simple translation of indices.

3.3.   FAC simulation.

As an illuminating example of its flexibility and ease of use, we show here how a global-grid unigrid code can be modified to account for local refinement. The procedure could hardly be simpler. Suppose for instance that a fine grid of mesh size $h = 1/(n+1)$ is desired in $\Omega_h = [0, \frac{1}{4}] \times [0, \frac{1}{4}]$, the lower left-hand corner of $\Omega = [0,1] \times [0,1]$. To convert the global-grid unigrid code to accommodate this local grid on the finest level, we would simply restrict relaxation there to Ω_h. In this case, all we would need to do is change the limits of the relevant do loop. For example, for the *unigrid* subroutine in Appendix A or B, after the definition of *ncoarse* we would simply add the statement

$$if\ (lvl.\ eq.\ nlvl.)\ ncoarse = (n-3)/4.$$

Certainly, other locations and shapes of the fine-grid patch would be more complicated to treat, but the only essential task here is to identify the indices that are interior to the patch location. This is vastly easier to do than to modify a conventional multigrid code to incorporate FAC.

An important warning is necessary here. It is certainly simple to modify unigrid to simulate FAC, but it is a bit more complicated to be sure that the code is working properly: by limiting the fine-grid directional iteration steps to a local region, we are virtually ensuring that the global-grid nodal equations cannot be solved, so no simple measure like the global-grid residual can be used to test for convergence. More specifically, since the unigrid direction vectors are defined on the global fine grid, but processing is restricted to the refinement region, then unless the solution to the unigrid equation looks like a grid $2h$ vector away from the refinement region, there is no chance to recover it. This means that the residual error norms cannot (and *should not*) tend to zero.

The problem here is that the unigrid residual error measures the ability of unigrid to solve the equation on the global fine grid, when it is the composite grid performance we should instead be measuring. One simple way around this difficulty that is useful for performance tests is to construct the source term, f, so that the solution is a legitimate composite grid function. This can be done by setting a vector equal to a specified or random linear combination of all grid $2h$ directions and local grid h directions, then defining f as the result of applying the operator to this vector. For the code listed in Appendix A, this can be done by adding the code segment listed in Appendix C to the *unigrid* subroutine. Of course, this initialization could also be accomplished in the driver routine that calls *unigrid*.

3.4.   Performance assessment.

In the design process of developing PML algorithms for a given application, it is important to know what performance characteristics to *expect*—and what to *accept*. While only in rare circumstances can theoretical analyses be used to predict performance, mode analysis (cf. [Stüben and Trottenberg 1982] and [Brandt 1984]) can often be used accurately for a priori estimation of worst-case convergence rates. It is not our intention to discuss the use or applicability of mode analysis, or any other performance prediction tool. Our aim here is rather to suggest a sensible a posteriori way to assess the performance of a given code, which should in any case be one of the first issues addressed in the algorithm design process. Actually, the real purpose of this section is to document two or three heuristic strategies that are commonly used by multigrid practitioners.

An important performance characteristic is the *worst-case algebraic convergence factor*. Assume that the code is designed to accept an initial approximation, then apply to it a specified number of cycles of unigrid (or multigrid, for that matter). Suppose that a certain algebraic error measure is recorded between each cycle. For example, the codes in Appendices A and B output the Euclidean norm of the residual error initially and after every (0,1) V-cycle. The *convergence factor* is then defined for a given cycle as the ratio of the error measure after and before that cycle. One of the simplest ways to estimate the *worst case* is to apply the code to a variety of initial guesses and to observe the largest ratio of successive error measures. However, for this process to be of any practical value, the following features of the test should be carefully considered:

> *Representative sample.* There must be some confidence that the initial guesses and subsequent iterates are sufficiently representative so that the worst-case condition is approximately achieved. For linear equations like the prototype Poisson problem, (1.1), a convenient strategy is to set the right-hand side to zero and use a representative set of initial guesses (chosen using a random number generator, for example). This has several advantages: there is no loss of generality because the initial guess is still arbitrary and, therefore, so is the error; the error is known because the exact solution is known (i.e., $u^{h*} = 0$); and, most importantly, this allows an almost unlimited number of iterations because machine precision does not prematurely stop the convergence process (near enough to a *nonzero* exact solution, the necessary correction is too small to affect the approximation). This latter property is important because, starting from virtually any initial guess, the *asymptotic convergence factor* (i.e., the convergence factor after many cycles) can be a good estimate of the worst case. In fact, the asymptotic factor almost always equals the worst-case energy factor for RML applied to a bilinear variation using a *self-adjoint* multigrid scheme, such as a (1,1) V-cycle where the point ordering for relaxation before coarsening is the reverse of the ordering afterwards. This results from the fact that iterating on the homogeneous equation is equivalent to applying the power method for finding the largest eigenvalue of the error propagation matrix, which for symmetric multigrid schemes is self-adjoint in the energy inner product; cf. [McCormick 1987, Chapter 5]. In this context, it may be useful to normalize the approximation in energy after each cycle; the normalization factor is then just the convergence factor for that particular cycle.

> *Mesh independence.* Multilevel methods have been developed that have

achieved optimal *algebraic complexity* for a large and growing number
of applications. The first ingredient of this optimality is a convergence
factor that is bounded less than one, independently of the size of the
problem. Of course, it would generally be unacceptable if this bound
were very near one, even if it did not degrade as the mesh size tended
to zero. Convergence factors of 0.2 or less are typical of multigrid basic
cycling schemes, except for certain very complicated applications (e.g.,
highly convective or inviscid fluid flows). In any case, performance tests
should be done over a range of values of the mesh size to observe the
dependency of the worst-case rate on this parameter.

Computational cost. Any convergent method can be made to do better than *any* given convergence factor—simply by defining one cycle to
consist of a sufficient number of iterations. The point here is that the
computational cost of each cycle must be accounted for in assessing performance. In the multigrid field, this cost is usually measured in *work
units* (WUs), which is basically the computational cost of evaluating
the relevant form (e.g., variation or equation) on the finest level. In the
case of local refinement, this should be taken to mean evaluation of the
form on the composite grid. For example, for the linear prototype, (1.1),
one WU is the cost of forming the vector of residuals, $\underline{r}^h = \underline{L}^h \underline{u}^h - \underline{f}^h$.
This basic unit can be thought of as measured in terms of the number
of floating point operations, or elapsed computer processing time (which
is useful for parallel machines). If the discretization leads to the usual
nearest neighbor stencil, then this cost for scalar computation should be
essentially proportional to the number of grid points. For example, for
our linear prototype on an $n \times n$ uniform grid using the discretization
given by (1.27) and (1.28), one WU is the cost of about eight additions,
one subtraction, one multiplication (by 8), and one division (by 3) for
each grid point, yielding a total cost equal to that of about $11n^2$ floating
point operations. Note that a (1,1) V-cycle for this problem uses two
relaxations on each level, where the number of points starting from the
finest are about $n^2, \frac{n^2}{4}, \frac{n^2}{16}, \cdots$. Since one Gauss-Seidel relaxation sweep
costs about one WU, then, counting the process of forming the residual
before coarsening as one WU, we obtain a cost in WUs bounded by

$$3 \sum_{\ell=0}^{\infty} 4^{-\ell} = 4.$$

The total cost of a (1,1) V-cycle, including the relatively inexpensive
intergrid transfers, is therefore less than 5 WUs.

The linear prototype can be treated by multigrid schemes based on a cycling
strategy that will in worst case *reduce the algebraic error by a decimal place at
a cost of just a few work units.* Using this as a guide, it seems reasonable to
attempt to achieve similar performance for other applications. This "optimal"

performance provides somewhat of a rule-of-thumb for multigrid practitioners. However, there are instances where the complexity is significantly greater, though still optimal (e.g., inviscid fluid flow where worst-case convergence factors are often 0.9 or worse). Of course, there are many problems where such optimal algebraic complexity has not yet been achieved by multigrid processing.

Assessment of the algebraic performance of basic multilevel cycling strategies is often not enough. Since the real goal of computation is presumably the numerical solution of the PDE, it is important to evaluate the effectiveness of the overall scheme in achieving that goal. In other words, performance tests should measure *actual complexity*, which assesses what actual errors are obtained by the scheme, and at what costs. Typically for such tests, V-cycles are inadequate, and we must turn to the more sophisticated FMG-cycles (cf. [Brandt 1984]).

To be more specific, a seemingly reasonable premise is that the real objective of the numerical method is to obtain an *acceptable bound on the actual error*. First of all, it is almost useless to have a small error without knowing that it is small; so it is really the *bound* that counts, not the error itself. Second, estimates of the algebraic error are not enough: we must know that the nodal vector produced by the scheme accurately represents the PDE solution, so it is the *actual* error that we are after. Letting $\| \cdot \|$ denote some norm on H_1, then our computational goal is to ensure that the final approximation has a suitably small actual error, e:

$$\|e\| \leq \varepsilon,$$

where $\varepsilon > 0$ is some prescribed tolerance. Remember that the actual error is the sum of the discretization and algebraic errors:

$$e = e^* + e^h.$$

Then it seems reasonable to achieve our computational goal by ensuring both

$$(3.6) \qquad \|e^*\| < \frac{\varepsilon}{2}$$

and

$$(3.7) \qquad \|e^h\| < \frac{\varepsilon}{2}.$$

To require one to be significantly smaller than the other would be computationally wasteful. (Actually, a carefully tuned objective would take into account the problem dimension, the order of discretization accuracy, and the cost and convergence factors for the grid h and grid $h/2$ V-cycles; cf. [Brandt 1984, §7.2]. Such an analysis might suggest in some cases that the algebraic error tolerance be somewhat smaller than the actual error tolerance. However, it is seldom that these tolerances should differ by more than an order of magnitude, and, even in such cases, the principle underlying the following argument is basically the same.)

Now (3.6) dictates the size of the mesh size, h. For example, if a bound on the discretization error is known to behave like ch^2, then (3.6) is achieved by choosing

$$(3.8) \qquad\qquad h = \left(\frac{\varepsilon}{2c}\right)^{1/2}.$$

(If c is unknown, it can be estimated by assessing typical discretization errors as described in Chapter 1, §1.7. Note that h might be chosen more conveniently as, say, the negative power of 2 closest to satisfying (3.8).) Assuming for discussion that h is chosen in this way, then (3.7) can be rewritten as

$$(3.9) \qquad\qquad \|e^h\| \le ch^2.$$

This expresses the convergence criterion that can be difficult for algebraic solvers to achieve: given an initial guess (e.g., $u^h = 0$) with error presumably of order 1, then even a solver with optimal algebraic convergence factors (bounded less than one independent of h) must perform on the order of $\log(1/h)$ iterations to be sure that (3.9) is satisfied. FMG methods circumvent this trouble to achieve full optimality by using coarser levels first to determine a starting approximation for the V-cycles that is already fairly accurate: if u^{2h} satisfies the $2h$ version of (3.9), then its actual error can be no worse than $8ch^2$ (the sum of its discretization and algebraic error bounds), so it must approximate u^{h^*} by an error no larger than $9ch^2$; thus, (3.9) can be achieved by just one V-cycle with a convergence factor bounded by $1/9$. Thinking this through recursively leads to the FMG scheme that cycles from the coarsest to the finest grid, using a V-cycle on each level applied to the approximation computed on the coarser level. In a unigrid framework, using $(0,1)$ V-cycles, this FMG scheme is written, analogous to (3.4), as follows:

For $\mu = 2, 3, \cdots, p$:

　For $\nu = 1, 2, \cdots, \mu$:

　　For $\ell = 1, 2, \cdots, m^h\nu$:

$$u^h \leftarrow D^h(u^h; d^{h_\nu}_{(\ell)}).$$

Actually, $(0,1)$ V-cycles are usually not sufficiently effective to reduce the error by $1/9$. $(1,1)$ V-cycles or $(2,1)$ V-cycles are common in practice, but the best choice must be determined by numerical or theoretical analyses, or both.

To summarize, thorough tests of the performance of a multigrid code should carefully estimate worst-case algebraic convergence factors of the basic cycling scheme and actual errors left in the results produced by the overall scheme. The tests should include a spectrum of mesh sizes and problem data (e.g., source terms and boundary conditions), and the estimates should be measured against the computational costs of achieving the desired accuracy.

3.5. Caveats.

Unigrid can provide a very useful tool for design of multigrid algorithms. It is especially effective for experimentation with several algorithm variants, including different cycling strategies, local refinement, and various forms of relaxation. However, it has several limitations that should always be kept in mind.

The major disadvantage of the unigrid method is its efficiency. A little experience with unigrid should convince anyone that it will seldom if ever be useful as a technique in production software. It is important to understand this slowness beforehand so that it never comes as an imposing disappointment. For example, the very long processing times stemming from unigrid inefficiency create a practical limitation on problem size that may be too restrictive to give full confidence in the multigrid performance that it is simulating.

Another limitation of unigrid is its restriction to projection-based relaxation and coarsening processes. It is awkward if not practically impossible to try fundamentally different strategies in the unigrid context.

The commitment to point relaxation methods would seem just as strong because unigrid corrections are based on single vectors, not subspaces. Actually, though, this is not quite the case. In fact, block relaxation can often be approximated by using either many unigrid relaxations within the block, or special unigrid directions that are designed to simulate multigrid within the blocks. For example, line relaxation on a given level can be accomplished by an inner unigrid scheme that involves one-dimensional "coarsening" of the directions associated with each line. (Note, for example, that the coarsest direction for relaxation on a level $2h$ y line is a skinny hat function that has support of width $4h$ in the x direction but extends up to both boundaries in the y direction. See Figure 3.2.)

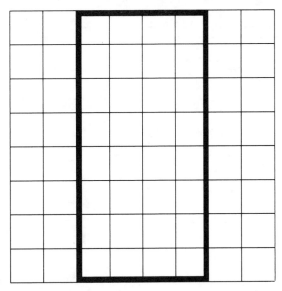

FIG. 3.2. *Support for coarsest-level hat function for y-line relaxation on level* $2h$.

A third major difficulty with unigrid is that it circumvents the question of coarse-level realizability. Of course, this feature is just what makes unigrid so useful because it also avoids the need to construct this realization. However, for most studies, it will be essential to certify that the final algorithm design can be implemented in a code that exhibits the usual multigrid complexity. This can often be proved analytically, but perhaps the most convincing argument is to produce the code itself. So, in later stages of development, it is often best to abandon unigrid in favor of a true multigrid implementation.

CHAPTER **4**

Paradigms

We have carried along two important prototype problems in an attempt to illustrate the basic principles developed in the first three chapters. These principles and the techniques they induce have been very well developed and analyzed in the open literature for the linear problem, (1.1). To a lesser extent, this is also true for the eigenvalue problem, (1.2). However, these prototypes alone do not justify the effort which we have just undertaken to develop these principles into a general methodology. We therefore devote this chapter to exploring several other specific applications of the general approach.

The paradigms described in the following sections are in widely varying states of development. While none is even close to being fully explored, some have progressed enough to show encouraging potential. Others, however, are currently just ideas. In any case, each of these examples will be presented in a fairly succinct way, with no numerical results, no theory, and little in the way of concrete motivation. We will attempt to present these paradigms in a clean and mostly formal way, choosing the simplest model problems and the most natural PML schemes, and being vague about the relevant function spaces and other constructs.

There are several risks taken here, not the least of which is that we are overlooking subtleties that threaten the soundness of the whole approach. In fact, while this monograph was being written, we discovered an essential difficulty with our initial formulation of the parameter estimation problem. But, instead of describing our new approach to this problem, we chose to stay with the old paradigm because it provides an interesting example of how multigrid can expose trouble with existing mathematical models. *After all, we are developing paradigms, with the sole central aim of illustrating the potential for the general methodology to apply to a wide variety of scientific problems; actual development of viable techniques, careful mathematical constructions, and overall proof of concept will have to wait for much more study.*

This chapter develops six paradigms, two each using Rayleigh-Ritz, Galerkin, and Petrov-Galerkin projection schemes, presented in that order.

61

4.1. Rayleigh-Ritz 1: Parameter estimation.

PDEs are often found at the core of numerical simulation. In fact, the state of many physical processes can be described as the solution of a static or time-dependent boundary value problem. Unfortunately, while the form of the PDE may be known, frequently the coefficients are not. Thus, one of the major tasks of mathematical modeling is to experiment with the real physical process under a variety of imposed conditions, observe specific behavior of the state of the system, then use these observations to deduce the unknown coefficients.

A prime example of this *inverse* problem comes from electrical impedance tomography. The electrical voltage in a given conducting body is determined by imposed voltage or current fluxes on its surface. The voltage is known to obey an elliptic equation, where the coefficient represents the conductivity of the body. The central task in electrical impedance tomography is to estimate this coefficient by using observations from experiments that measure both the voltage and the current fluxes at the boundary, under varying conditions of state. Typically, a variety of flux conditions are imposed on the body's surface, and resulting voltages are observed there. The premise in inverse problems is that enough observations on the state of the system will make it possible to recover some of the details of the equations that govern it. In the case of electrical impedance tomography, these details are the interior values of conductivity, which yield an image of the object under study.

We will use a special formulation of the inverse problem of this type that is applicable to two-dimensional steady-state elliptic equations. (See [McCormick and Wade 1991].) This form, which was introduced in [Kohn and Vogelius 1987], uses the variation

$$(4.1) \qquad R(a; u_{(.)}; v_{(.)}) \equiv \int_{\Omega} \left(a \sum_{k=1}^{m} \|\nabla u_{(k)}\|^2 + a^{-1} \sum_{k=1}^{m} \|\nabla v_{(k)}\|^2 \right) dz,$$

where $\Omega \subset \Re^2$ is the unit square, $\|\cdot\|$ is the L_2 norm on $\Re^2$, a is the conductivity parameter, and $u_{(.)}$ and $v_{(.)}$ are the state variables. a, $u_{(.)}$, and $v_{(.)}$ are assumed to lie in appropriate Hilbert spaces of functions defined on Ω. (We use $u_{(.)}$ to denote $u_{(k)}$, $k = 1, 2, \cdots, m$, collectively, and similarly for $v_{(.)}$ and other quantities that follow.) Let $u_{(k)}^0$ and $v_{(k)}^0$ denote functions defined on the boundary, Γ, of Ω, $1 \le k \le m$. Then the problem that we consider here is to minimize R in (4.1) subject to the restrictions

$$a > 0$$

and

$$(4.2) \qquad u_{(k)} = u_{(k)}^0 \text{ and } v_{(k)} = v_{(k)}^0 \text{ on } \Gamma,$$

$1 \le k \le m$.

This is actually a special formulation of the parameter estimation problem, which arises in the following way. Consider for simplicity the special case of one observation, $m = 1$. The steady-state elliptic equation is given by

$$(4.3) \qquad -\nabla \cdot a\nabla u = 0,$$

which can be rewritten as the system

$$a \nabla u + \underline{w} = 0$$

(4.4)
$$\nabla \cdot \underline{w} = 0.$$

Note that the *continuity* condition $\nabla \cdot \underline{w} = 0$ implies the existence of a *stream function* v such that

$$\underline{w} = \begin{pmatrix} -v_y \\ v_x \end{pmatrix}.$$

Thus, the variation

(4.5)
$$\int_\Omega \left\| a^{1/2} \nabla u + a^{-1/2} \begin{pmatrix} -v_y \\ v_x \end{pmatrix} \right\|^2 dz$$

measures how well u and $\underline{w} = \begin{pmatrix} -v_y \\ v_x \end{pmatrix}$ satisfy the first equation in (4.4). Now the parameter estimation problem is posed as one of observing the boundary data for a solution, u and $\underline{w} = \begin{pmatrix} -v_y \\ v_x \end{pmatrix}$, of (4.4) and determining the parameter $a > 0$ that best fits these observations in the sense that (4.5) is minimized over a, u, and v. The observed boundary values include Dirichlet and Neumann data for u, which translates to Dirichlet data for u and $\underline{w}$. Since data for $\underline{w}$ can be integrated on Γ to form data for v, then the parameter estimation problem becomes one of minimizing the functional in (4.5) subject to given Dirichlet data on u and v. This is seen to be equivalent to the formulation above by noting that the difference between the functionals in (4.1) (with $m = 1$) and (4.5) is a quantity that, using integration by parts, consists of integrals of terms involving only the boundary values of u and v, and so is constant with respect to minimization.

Returning to (4.1)–(4.2), note that this variational problem is not strictly of the form we assumed in the earlier chapters because the boundary conditions are inhomogeneous. This is of course necessary for the problem even to make sense because homogeneous boundary data imply that all observed solutions are zero, which does nothing to characterize the parameter. We could use the standard transformation that replaces $u_{(\cdot)}$ by $u_{(\cdot)} + w_{(\cdot)}$, where $w_{(\cdot)}$ is chosen to satisfy the boundary condition, and similarly for $v_{(\cdot)}$. However, we prefer for convenience to stay with the original formulation, which in any case does not prevent us from applying the multilevel projection methodology.

To form the discretization of (4.1)–(4.2) on a uniform grid, Ω^h, let S_a^h be the space of functions that are piecewise constant, and let S^h be the space of continuous functions that are piecewise bilinear, with respect to the cells of Ω^h. The discrete functional is defined by

$$R^h(a^h; u_{(\cdot)}^h; v_{(\cdot)}^h) \equiv R(P_{S_a^h} a^h; P_{S^h} u_{(\cdot)}^h; P_{S^h} v_{(\cdot)}^h),$$

$a^h \in S_a^h$, $u_{(k)}^h \in S^h$, and $v_{(k)}^h \in S^h$, $1 \leq k \leq m$. The computational goal is to minimize this functional over its arguments subject to the constraints

$$a > 0$$

and

$$u_{(k)}^h = u_{(k)}^{0^h} \quad \text{and} \quad v_{(k)}^h = v_{(k)}^{0^h} \quad \text{on } \Gamma,$$

where $u_{(k)}^{0^h}$ and $v_{(k)}^{0^h}$ are the respective interpolants of $u_{(k)}^0$ and $v_{(k)}^0$ on the boundary, Γ^h, of Ω^h, $1 \leq k \leq m$. (By u^{0^h} being the interpolant of u^0 on Γ^h we mean that u^{0^h} is the unique function that is continuous piecewise linear with respect to the cells of Γ^h and agrees with u^0 at its nodes.)

To determine the discrete representation of R^h in terms of the parameter cell values and state nodal values, first note that the integral in (4.5) can be written as a sum of integrals over the cells of Ω^h, on each of which a is constant. Thus, a simple calculation based on writing the state variables as bilinear functions of their nodal values yields the expression

$$R^h(a^h; u_{(\cdot)}^h; v_{(\cdot)}^h) = \sum_{k=1}^{m} \sum_{i,j=0}^{n} \left(a_{i+1/2\,j+1/2}^h \, \alpha_{i+1/2\,j+1/2}^h(u_{(k)}^h) \right.$$

(4.6)
$$\left. + (a_{i+1/2\,j+1/2}^h)^{-1} \alpha_{i+1/2\,j+1/2}^h(v_{(k)}^h) \right).$$

Here we have used the quantity $\alpha_{i+1/2\,j+1/2}^h(u^h)$, which is defined in terms of the differences $\delta_{i+1/2\,j}^h(u^h) = u_{i+1\,j}^h - u_{i\,j}^h$, $\delta_{i+1/2\,j+1}^h(u^h) = u_{i+1\,j+1}^h - u_{i\,j+1}^h$, $\delta_{i\,j+1/2}^h(u^h) = u_{i\,j+1}^h - u_{i\,j}^h$, and $\delta_{i+1\,j+1/2}^h(u^h) = u_{i+1\,j+1}^h - u_{i+1\,j}^h$ by

$$\alpha_{i+1/2\,j+1/2}^h(u^h) = \frac{1}{3} \left((\delta_{i+1/2\,j}^h(u^h))^2 + (\delta_{i+1/2\,j}^h(u^h))(\delta_{i+1/2\,j+1}^h(u^h)) \right.$$
$$+ (\delta_{i+1/2\,j+1}^h(u^h))^2 + (\delta_{i\,j+1/2}^h(u^h))^2$$
$$\left. + (\delta_{i\,j+1/2}^h(u^h))(\delta_{i+1\,j+1/2}^h(u^h)) + (\delta_{i+1\,j+1/2}^h(u^h))^2 \right),$$

$0 \leq i, j \leq n$. (We use half indices for cell centers, $n \geq 1$ to represent the number of interior grid lines in both coordinate directions of Ω^h, and $a_{i+1/2\,j+1/2}^h$ to denote the value of a^h on cell $i + 1/2\,j + 1/2$, $0 \leq i, j \leq n$.) Note that (4.6) shows that the discretization of R is realizable.

The simplest form of relaxation on (4.6) is fairly straightforward. State relaxation is conventional because, for fixed a, R^h is just a sum of discrete quadratic forms for $2m$ homogeneous diffusion equations. Thus, relaxation of one of the state variables is just standard point Gauss-Seidel applied to the associated homogeneous system of state. Note that individual state variables can be processed independently of the others. Parameter relaxation is even simpler. In fact, for fixed state, since there is no interaction between the cells directly via the parameter, then (4.6) can be minimized exactly in one step by the formula

$$(4.7) \qquad a^h_{i+1/2\,j+1/2} = \left(\frac{\displaystyle\sum_{k=1}^{m} \alpha^h_{i+1/2\,j+1/2}(v^h_{(k)})}{\displaystyle\sum_{k=1}^{m} \alpha^h_{i+1/2\,j+1/2}(u^h_{(k)})} \right)^{1/2},$$

$0 \le i,\, j \le n$. A sensible relaxation strategy for (4.6) is to first perform point Gauss-Seidel on each of the state variables, holding the parameter fixed, then finish with a relaxation on the parameter via (4.7). This process constitutes one full relaxation sweep for minimizing (4.6). It is also possible to use other more complicated relaxation strategies that involve interaction within the state variables and between the states and the parameter.

The coarsening process is based on the subspace, S_0^{2h}, of S^h of functions which are continuous piecewise bilinear with respect to the cells of $\Omega^{2h} \subset \Omega^h$, and which are zero on $\partial\Omega$. Here we assume that $n > 1$ is odd so that Ω^{2h} is a conforming uniform subgrid of dimension $(n-1)/2 \times (n-1)/2$. We also define S_a^{2h} as the subspace of S_a^h of piecewise constant functions with respect to the cells of Ω^{2h}. Then the natural coarse-grid correction functional is given by

$$(4.8) \qquad R^{2h}(a^{2h}; u^{2h}_{(\cdot)}; v^{2h}_{(\cdot)}) \equiv R^h(a^h \cdot a^{2h}; u^h_{(\cdot)} + u^{2h}_{(\cdot)}; v^h_{(\cdot)} + v^{2h}_{(\cdot)}),$$

$a^{2h} \in S_a^{2h}$, $u^{2h}_{(k)} \in S_0^{2h}$, and $v^{2h}_{(k)} \in S_0^{2h}$, $1 \le k \le (n-1)/2$. For the state variables, this uses the usual *additive* correction scheme. However, the parameter is treated by a *multiplicative* correction, which is more natural and which, as we now show, enables the correction to be coarse-grid realizable.

To argue heuristically that R^{2h} is expressible in terms of a reasonable number of elementary Euclidean operations on Ω^{2h} (the details of a constructive argument would require cumbersome notation and too much detail), consider a typical component of R^h, which we write as

$$Q^h(b^h; w^h) \equiv \int_\Omega b^h (w^h_x)^2 \, dz.$$

Here, $b^h \in S_a^h$ represents either a^h or $(a^h)^{-1}$ and $w^h \in S^h$ (with $w^h = w^{0h}$ on Γ) represents one of the state variables. Note that R^h can be written as a sum of terms of this type and of the type $\int_\Omega b^h (w^h_y)^2 \, dz$, which is treated analogously. Now the component of R^{2h} corresponding to this term is given by

$$Q^{2h}(b^{2h}; w^{2h}) \equiv Q^h(b^h \cdot b^{2h}; w^h + w^{2h}),$$

$b^{2h} \in S_a^{2h}$ and $w^{2h} \in S_0^{2h}$. This gives

$$Q^{2h}(b^{2h}; w^{2h}) = \\ \sum_{i,j=0}^{(n-1)/2} b^{2h}_{i+1/2\,j+1/2} \int_{\Omega^{2h}_{i+1/2\,j+1/2}} (b^h(w^h_x)^2 + (2b^h w^h)w^{2h}_x + b^h(w^{2h}_x)^2) \, dz,$$

where $\Omega^{2h}_{i+1/2\,j+1/2}$ is cell $i+1/2 \quad j+1/2$ of Ω^{2h}, $0 \leq i,\, j \leq (n-1)/2$. This relation for Q^{2h} shows that we may restrict our attention to a typical coarse-grid cell, $\Omega^{2h}_{i+1/2\,j+1/2}$, which corresponds to four fine-grid cells, $\Omega^{h}_{2i+\mu\,2j+\nu}$, $\mu, \nu = \frac{1}{2}, \frac{3}{2}$. Close inspection shows that R^{2h} is just a generalized form of the fine-grid variation, R^h, and that it is indeed expressible in terms of a reasonable number of Euclidean operations on Ω^{2h}. In fact, the first term in Q^{2h}, given by the expression

$$\sum_{i,j=0}^{(n-1)/2} b^{2h}_{i+1/2\,j+1/2} \int_{\Omega^{2h}_{i+1/2\,j+1/2}} b^h (w^h_x)^2\, dz,$$

is a linear term in b^{2h}, which R^h does not have. Note that the integral term in this expression is just the cell $\Omega^{2h}_{i+1/2\,j+1/2}$ component of the fine-grid "residual" $R^h(a^h; u^h_{(\cdot)}; w^h_{(\cdot)})$. The second term in Q^{2h}, given by

$$\sum_{i,j=0}^{(n-1)/2} 2b^{2h}_{i+1/2\,j+1/2} \int_{\Omega^{2h}_{i+1/2\,j+1/2}} (b^h w^h) w^{2h}_x\, dz,$$

is bilinear with respect to (b^{2h}, w^{2h}), which again has no counterpart in R^h since it corresponds to a nonzero source term for the elliptic equations of state. Finally, the third term in Q^{2h}, given by

$$\sum_{i,j=0}^{(n-1)/2} b^{2h}_{i+1/2\,j+1/2} \int_{\Omega^{2h}_{i+1/2\,j+1/2}} b^h (w^{2h}_x)^2\, dz,$$

is analogous to the individual cell terms in R^h, except for the integral weighting term b^h. In any event, each of these terms can be easily evaluated by performing operations on grid $2h$ involving quantities that are computed and transferred from grid h at the beginning of the coarsening cycle. This gives the usual sense of grid $2h$ realizability. It should also be clear that this process is recursive: the coarsening of Q^{2h} yields a grid $4h$ form that *is* analogous to Q^{2h}, so this heuristic argument extends to all coarser levels.

To complete the formal development, we need to be sure that relaxation can be easily performed on all coarser levels. This is clear for parameter relaxation because the parameter appears in R^{2h} in precisely the same form as it appears in R^h, so a relaxation formula analogous to (4.7) holds on coarser levels. For the state, it is enough to realize that each coordinate variable appears as a quadratic polynomial in R^{2h}. In fact, state relaxation is just Gauss-Seidel applied to the corresponding inhomogeneous elliptic equation with a source that arises from linear state terms in R^{2h}.

Unfortunately, while this formal development seems very natural, our numerical experience with the RML scheme based on the functional, R, defined in (4.1) has been disappointing. The trouble here probably stems not from the method, but from the functional itself: R seems to have such a degree of singularity in it that perhaps no natural approach will be able to determine its minimum effectively. This essential difficulty with R comes somewhat as a surprise because it has been apparently overlooked by those who use it in practice, despite an otherwise careful analysis (cf. [Kohn and Vogelius 1985 and 1987]). This paradigm is therefore all the more interesting as an example of how poor multigrid performance can expose trouble with the basic approach.

To see this trouble clearly, we will consider a functional, defined only in terms of the state variables $u_{(\cdot)}$ and $v_{(\cdot)}$, that is equivalent to R, then assume we are so close to the minimum that the new functional is quadratic in state. First notice that if we are at a minimum, then the parameter in (4.1) must satisfy

$$a = \left(\frac{\sum\limits_{k=1}^{m} \|\nabla v_{(k)}\|^2}{\sum\limits_{k=1}^{m} \|\nabla u_{(k)}\|^2} \right)^{1/2} \qquad \text{(pointwise)}.$$

This shows that minimizing R in (4.1) is equivalent to minimizing the *parameter-eliminated functional*

$$\tilde{R}(u;v) = \int_{\Omega} \left(\sum_{k=1}^{m} \|\nabla u_{(k)}\|^2 \right)^{1/2} \left(\sum_{k=1}^{m} \|\nabla v_{(k)}\|^2 \right)^{1/2} dz.$$

Here we drop the subscript $(\cdot)$ for convenience, so u and v now stand for the respective $u_{(k)}$ and $v_{(k)}$ collectively. For simplification, we abuse the notation further and rewrite $\tilde{R}$ as

$$\tilde{R}(u;v) = \int_{\Omega} \|\nabla u\| \cdot \|\nabla v\| \, dz,$$

where u and v are now thought of as block vector functions with respective entries $u_{(k)}$ and $v_{(k)}$, and the pointwise norms are given by

$$\|\nabla u\| = \left(\sum_{k=1}^{m} \|\nabla u_{(k)}\|^2 \right)^{1/2} \quad \text{and} \quad \|\nabla v\| = \left(\sum_{k=1}^{m} \|\nabla v_{(k)}\|^2 \right)^{1/2}.$$

Taking the Fréchet derivative of $\tilde{R}(u;v)$ in the direction $(p;q)$, then taking the Fréchet derivative of the result again in the direction $(p;q)$, yields the Hessian form

(4.9)
$$\tilde{R}''(u;v)(p;q)^2 = \int_{\Omega} \|\nabla u\| \cdot \|\nabla v\| [\|\overline{\nabla p}\|^2 + \|\overline{\nabla q}\|^2 - (\langle \overline{\nabla u}, \overline{\nabla p} \rangle - \langle \overline{\nabla v}, \overline{\nabla q} \rangle)^2] \, dz,$$

where we use the notation (note that ∇p and ∇q are "normalized" by $\|\nabla u\|$ and $\|\nabla v\|$, respectively)

$$\overline{\nabla u} = \nabla u / \|\nabla u\|, \quad \overline{\nabla v} = \nabla v / \|\nabla v\|, \quad \overline{\nabla p} = \nabla p / \|\nabla u\|, \quad \overline{\nabla q} = \nabla q / \|\nabla v\|.$$

Now consider the one-observation case $u = u_{(1)} = x$ and $v = v_{(1)} = y$:

$$\tilde{R}''(u;v)(p;q)^2 = \int_\Omega [p_x^2 + p_y^2 + q_x^2 + q_y^2 - (p_x - q_y)^2]\, dz$$

$$= \int_\Omega [p_y^2 + q_x^2 + 2p_x q_y]\, dz$$

$$= \int_\Omega [p_y + q_x]^2\, dz,$$

where the last line follows from integration of the term $p_x q_y$ by parts and the fact that p and q are zero on $\partial\Omega$. Now the trouble is apparent: if we let

$$p(x,y) = \alpha'(x)\beta(y) \quad \text{and} \quad q(x,y) = -\alpha(x)\beta'(y),$$

where $\alpha = \alpha(x)$ and $\beta = \beta(y)$ are *any* (suitably smooth) functions whose values and derivatives are zero at $x, y = 0, 1$, then

$$\tilde{R}''(u;v)(p;q)^2 = 0\,!$$

For example, we could take

$$\alpha(x) = \sin^2 k\pi x \quad \text{and} \quad \beta(y) = \sin^2 \ell\pi y,$$

so that p and q are as oscillatory as we like, yet $(p;q)$ is a null-space component for $\tilde{R}''$. If we now consider the case of the two observations $u = \frac{1}{\sqrt{2}}\begin{pmatrix} x \\ y \end{pmatrix}$ and $v = \frac{1}{\sqrt{2}}\begin{pmatrix} y \\ x \end{pmatrix}$ and let $p = \begin{pmatrix} p_1 \\ p_2 \end{pmatrix}$ and let $q = \begin{pmatrix} q_1 \\ q_2 \end{pmatrix}$, then we obtain a similar result:

$$\tilde{R}''(u;v)(p;q)^2 = \int_\Omega \left[p_{1_x}^2 + p_{1_y}^2 + p_{2_x}^2 + p_{2_y}^2 + q_{1_x}^2 + q_{1_y}^2 + q_{2_x}^2 + q_{2_y}^2 \right.$$

$$\left. - \left(\frac{p_{1_x} + p_{2_y} - q_{1_y} - q_{2_x}}{\sqrt{2}} \right)^2 \right] dz,$$

which, with the choice $q_1 = -p_2$ and $q_2 = -p_1$ and using integration by parts, yields

$$\tilde{R}''(p;q)(u;b)^2 = 2 \int_\Omega (p_{1_y} - p_{2_x})^2\, dz;$$

this shows that null-space components for $\tilde{R}''$ include

$$p_1(x, y) = \alpha'(x)\beta(y), \qquad p_2(x, y) = \alpha(x)\beta'(y),$$

$$q_1(x, y) = -\alpha(x)\beta'(y), \qquad q_2(x, y) = -\alpha'(x)\beta(y),$$

where $\alpha = \alpha(x)$ and $\beta = \beta(y)$ are as above. We speculate that this is typical of the situation for the general case of $m > 1$ observations of general type.

Minimization of a functional with a Hessian that has an infinite-dimensional null space is not necessarily a problem. For example, it may be that the finite element space used to discretize the functional is in the orthogonal complement of the null space, in which case all of the usual discretization estimates and stability properties might be obtained in this complement. (The finite element space may also be allowed to contain null space components, provided the rest of the space is in its orthogonal complement.) Real trouble arises when the finite element space is arbitrarily close to the null space of the Hessian, without intersecting it. This means that the discrete Hessian has arbitrarily bad ellipticity (e.g., its condition number is much larger than $O(h^{-2})$) and that null-space components of the continuum Hessian are likely to contaminate critically the approximation of some or all of the well-posed components. This is the situation we face with the functional defined in (4.1). We have therefore chosen to abandon this formulation of the parameter estimation problem in favor of the more conventional approach of output least squares, which we are just beginning to explore.

As with most inverse problems, most forms of the parameter estimation problem are physically ill-posed. This is suggested by the fact that we are attempting to use boundary data to estimate quantities defined on the interior of the region. While it is true that knowledge of the exact Dirichlet and Neumann boundary values for a complete (and, therefore, infinite) set of observations may mathematically determine the parameter (cf. [Kohn and Vogelius 1985], which treats the function we defined in (4.1)), physical and computational constraints limit the number of observations, the accuracy in the data, and the resolution and accuracy of the discrete system. As a consequence, it is essential to examine how these limitations and the problem's ill-posedness affect the numerical approximations—and even how we are to interpret the results that are obtained. We must therefore be more specific about what is meant by a solution to the problem, and this additional characterization must be effectively incorporated into the numerical scheme. Unfortunately, many conventional methods for handling this ill-posedness take a symptomatic approach that does not account for the specific source of the trouble. For example, the functional might be regularized by adding a term that tends to eliminate oscillatory parameter components, even though *some* oscillatory aspects of the parameter can be accurately estimated using the original functional.

Based on an observation by Achi Brandt, we now suggest how the multilevel projection method can make parameter estimation problems well posed essentially by filtering out solution components that are not correctly determined, without harm to those that are. To do this, we must first characterize the well-posed components. Considering the physical model defined in (4.3), it is not difficult to see that any oscillatory behavior of the parameter has negligible effect on the state variables far from where this behavior occurs, relative to the scale of the oscillations. More specifically, if a is perturbed by a term that has zero mean but oscillates at a frequency of Δz in a region in Ω that contains the support of this term, then the change in any solution of $-\nabla a \nabla u = 0$ will be negligible at a distance of several Δz's away from this region. This suggests that the problem can be better posed by computing only those components of the parameter that are smooth in the interior of Ω and increasingly oscillatory near the boundary, Γ. The multilevel projection method can accomplish this by local refinement using fine-level grids that are increasingly local to Γ. See Figure 4.1 for an illustration of the grids we have in mind. The PML scheme developed for solving most parameter estimation models can easily incorporate local refinement of this type in the regularization and solution process.

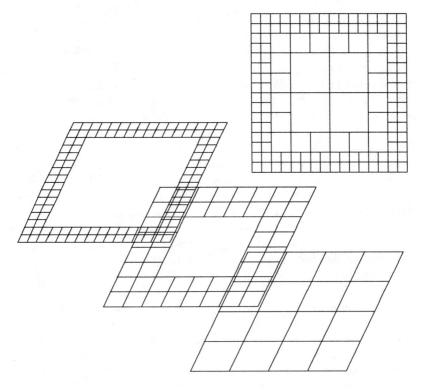

FIG. 4.1. *Sample composite grid and its subgrids for parameter estimation.*

4.2. Rayleigh-Ritz 2: Transport equations.

Particle transport plays a fundamental role in many scientific fields, including nuclear reactor design, radiation therapy, semiconductor device modeling, and satellite electronics shielding. The linear Bolzman transport equations govern the movement of neutrally charged particles. For steady-state, single group, isotropic scattering on the homogeneous semi-infinite slab $[-1, 1] \times (-\infty, \infty)^2$ in $\Re^3$, if we assume that the particle flux, u, is independent of y and z, then these equations simplify to the following:

$$(4.10) \qquad \mu u_x(x, \mu) + \sigma_t u(x, \mu) = \frac{\sigma_s}{2} \int_{-1}^{1} u(x, \nu) \, d\nu + f(x, \mu),$$

where σ_t and σ_s are constants (*probability density cross-sections*) satisfying $0 \leq \sigma_s \leq \sigma_t$, f is a given source, and the independent variables are the spatial distance, x, and the cosine of the angle from the x axis, μ. We consider (4.10) on the region $\Omega = [-1, 1] \times [-1, 1]$ subject to the following boundary conditions, which prescribe the flux of particles entering the slab:

$$(4.11) \qquad u(-1, \mu) = u^0(\mu), \quad 0 < \mu < 1, \quad u(1, \mu) = u^0(\mu), \quad -1 < \mu < 0.$$

For convenience, define the operator L by

$$(Lu)(x, \mu) = \mu u_x(x, \mu) + \sigma_t u(x, \mu) - \frac{\sigma_s}{2} \int_{-1}^{1} u(x, \nu) \, d\nu.$$

To develop the least squares formulation of (4.10)–(4.11), let H be an appropriate space of functions defined on Ω (e.g., $H = H^1[-1, 1] \times L_2[-1, 1]$) and let $\| \cdot \|$ be the L_2 norm for functions defined on Ω. Then the variation $R : H \to \Re$ is defined by

$$(4.12) \qquad R(u) = \|Lu - f\|^2,$$

$u \in H$.

While a least squares formulation is ill-advised for many PDEs, it seems natural here because it essentially converts a first-order equation to a variational principle for a well-behaved second-order equation. Another compelling attribute of this formulation is that, under the assumption that u is independent of μ, the resulting variation reduces to the energy functional for a two-point boundary value problem with an equation of the form

$$-u_{xx} + (\sigma_t - \sigma_s)^2 u = g.$$

The discretization of R is based on the subspace, $S^h \subset H$, of functions that are piecewise constant in μ and continuous piecewise linear in x with respect to the cells of a uniform grid, Ω^h, on Ω. (The mesh sizes of Ω^h in x and μ could, and no doubt should, be different. Moreover, it is perhaps more common to use other spatial discretizations, cf. [Manteuffel et al. 1990]. However, the assumptions here are only for simplicity and the following development carries over easily to more practical cases.) Note that the value of $u^h \in S^h$ in a cell of

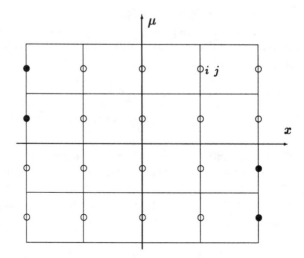

FIG. 4.2. *Uniform grid for transport (● signifies a Dirichlet boundary point, ○ signifies an interior or free boundary point, a cell is defined as a square bounded by grid lines).*

Ω^h is determined by its values at the midpoints of the cell's vertical sides, which we label u_{ij}^h, $i = 0, \pm 1, \cdots, \pm n$, $j = \pm 1, \pm 2, \cdots, \pm n$, where n is the number of cells of Ω^h in each positive coordinate direction. See Figure 4.2. In fact,

$$(4.13) \qquad u^h((i + s)h, (j + t)h) = (1 - s)u_{ij}^h + su_{i+1\,j}^h,$$

$0 \le s \le 1$, $-\frac{1}{2} < t < \frac{1}{2}$, $-n \le i \le n - 1$, and $j = \pm 1, \pm 2, \cdots, \pm n$.

The discrete variation, $R^h(u^h) \equiv R(P_{S^h} u^h)$, $u^h \in S^h$, can be expressed in terms of the nodal values of u^h as follows. First note that

$$R^h(u^h)(x, \mu) =$$

$$\int_{-1}^{1} \int_{-1}^{1} [\mu u_x^h(x, \mu) + \sigma_t u^h(x, \mu) - \frac{\sigma_s}{2} \int_{-1}^{1} u(x, \nu)\, d\nu - f(x, \mu)]^2 \, dx \, d\mu.$$

Clearly, $R^h(u^h)$ can be written as a sum of integrals over the cells of Ω^h. Let $i \in \{-n, -n + 1, \cdots, n - 1\}$ and $j \in \{\pm 1, \pm 2, \cdots, \pm n\}$ be fixed. Then the term in $R^h(u^h)$ due to cell $i + \frac{1}{2}\, j$ is

$$\delta_{i+1/2\,j}^h \equiv \int_{x_{i+1/2}-h/2}^{x_{i+1/2}+h/2} \int_{\mu_j-h/2}^{\mu_j+h/2} \Big[\mu u_x^h(x, \mu) + \sigma_t u^h(x, \mu)$$

$$- \frac{\sigma_s}{2} \int_{-1}^{1} u^h(x, \nu)\, d\nu - f(x, \mu) \Big]^2 \, dx \, d\mu.$$

Here, $x_{i+1/2} = (i + \frac{1}{2})h$ and $\mu_j = \nu_j h$, where $\nu_j = \begin{cases} j - \frac{1}{2} & j > 0 \\ j + \frac{1}{2} & j < 0 \end{cases}$. Translating the (x, μ) origin to $(x_{i+1/2}, \mu_j)$, assuming that f is piecewise constant on cells of Ω^h, and letting $f^h_{i+1/2\,j}$ denote the value of f on cell $i + \frac{1}{2}\,j$, then by (4.13) we have

$$
\begin{aligned}
\delta^h_{i+1/2\,j} = \int_{-h/2}^{h/2} \int_{-h/2}^{h/2} & \Bigg\{ (\mu_j + \mu) \left(\frac{u^h_{i+1\,j} - u^h_{i\,j}}{h} \right) \\
& + \sigma_t \left(\frac{u^h_{i\,j} + u^h_{i+1\,j}}{2} \right) + \sigma_t \left(\frac{u^h_{i+1\,j} - u^h_{i\,j}}{h} \right) x \\
& - \frac{h}{2} \sum_{k=\pm 1}^{\pm n} \left(\sigma_s \left(\frac{u^h_{i\,k} + u^h_{i+1\,k}}{2} \right) + \sigma_s \left(\frac{u^h_{i+1\,k} - u^h_{i\,k}}{h} \right) x \right) - f^h_{i+1/2\,j} \Bigg\}^2 dx\,d\mu \\
= \int_{-h/2}^{h/2} \int_{-h/2}^{h/2} & \Bigg\{ \left[\frac{\mu_j}{h}(u^h_{i+1\,j} - u^h_{i\,j}) + \frac{\sigma_t}{2}(u^h_{i\,j} + u^h_{i+1\,j}) \right. \\
& \left. - \frac{h}{2} \sum_{k=\pm 1}^{\pm n} \frac{\sigma_s}{2}(u^h_{i\,k} + u^h_{i+1\,k}) - f^h_{i+1/2\,j} \right] \\
& + \left[\sigma_t(u^h_{i+1\,j} - u^h_{i\,j}) - \frac{h}{2} \sum_{k=\pm 1}^{\pm n} \sigma_s(u^h_{i+1\,k} - u^h_{i\,k}) \right] \frac{x}{h} \\
& + [u^h_{i+1\,j} - u^h_{i\,j}]\frac{\mu}{h} \Bigg\}^2 dx\,d\mu \\
= h^2 \Bigg\{ & \left[\nu_j(u^h_{i+1\,j} - u^h_{i\,j}) + \frac{\sigma_t}{2}(u^h_{i\,j} + u^h_{i+1\,j}) \right. \\
& \left. - \frac{h}{2} \sum_{k=\pm 1}^{\pm n} \frac{\sigma_s}{2}(u^h_{i\,k} + u^h_{i+1\,k}) - f^h_{i+1/2\,j} \right]^2 \\
& + \frac{1}{12} \left[\sigma_t(u^h_{i+1\,j} - u^h_{i\,j}) - \frac{h}{2} \sum_{k=\pm 1}^{\pm n} \sigma_s(u^h_{i+1\,k} - u^h_{i\,k}) \right]^2 + \frac{1}{12}[u^h_{i+1\,j} - u^h_{i\,j}]^2 \Bigg\}.
\end{aligned}
$$

Hence, the full expression for $R^h(u^h)$ is

$$
\begin{aligned}
R^h(u^h) = h^2 \sum_{i=-n}^{n-1} \sum_{j=\pm 1}^{\pm n} & \Bigg\{ \left[\nu_j(u^h_{i+1\,j} - u^h_{i\,j}) + \frac{\sigma_t}{2}(u^h_{i\,j} + u^h_{i+1\,j}) \right. \\
& \left. - \frac{h}{2} \sum_{k=\pm 1}^{\pm n} \frac{\sigma_s}{2}(u^h_{i\,k} + u^h_{i+1\,k}) - f^h_{i+1/2\,j} \right]^2 \\
& + \frac{1}{12} \left[\sigma_t(u^h_{i+1\,j} - u^h_{i\,j}) - \frac{h}{2} \sum_{k=\pm 1}^{\pm n} \sigma_s(u^h_{i+1\,k} - u^h_{i\,k}) \right]^2 \\
& + \frac{1}{12}[u^h_{i+1\,j} - u^h_{i\,j}]^2 \Bigg\}.
\end{aligned}
$$

(4.14)

Transport problems in general suggest a wide variety of relaxation schemes. The effectiveness of a given scheme in a multilevel context will depend on the particular character of the problem in question. For the specific category of transport problems considered here, the character that probably most influences performance are the constants σ_t and σ_s. In the extreme case $\sigma_t = \sigma_s = 0$, minimizing $R^h(u^h)$ reduces to $2n$ independent discrete initial value problems in one dimension. This suggests that problems with small $\sigma_t h$ ($\geq \sigma_s h \geq 0$) would be effectively treated by point Gauss-Seidel with the ordering along horizontal lines of grid cells, beginning for each line at its Dirichlet boundary (left end for positive μ, right end for negative μ). At the other extreme where $\sigma_t h$ is very large, if σ_s is small compared to σ_t, then $R^h(u^h)$ is dominated by the influence of the mass matrix arising from the operator term $\sigma_t I$ in L. (I is the identity operator.) In this case, point Gauss-Seidel using any ordering and by itself, *without* multileveling, should be effective. When $\sigma_t h$ is large and $\sigma_s h$ is (nearly) equal to it, $R^h(u^h)$ is dominated by the influence of a matrix arising from the (nearly) singular operator $\sigma_t I - \sigma_s P$ in L, where P is defined by $Pu(x,\mu) = \frac{1}{2} \int_{-1}^{1} u(x,\nu) \, d\nu$. This suggests a μ-*line* relaxation scheme, where, for each i in turn, the u_{ij}^h are changed over all $j \in \{\pm1, \pm2, \cdots, \pm n\}$ so that the resulting $R^h(u^h)$ is as small as possible. This relaxation scheme is particularly easy to perform since the resulting linear equation that defines this optimal change involves the $2n \times 2n$ rank one matrix consisting of all ones. The interaction between different values of μ occurs only through this matrix, so the linear equation defining μ-line relaxation is especially easy to solve. Taking all of these heuristics into consideration, a likely candidate for general σ_t and σ_s is μ-line relaxation performed alternately from left to right and right to left.

The coarsening of R^h is straightforward. In fact, L is linear, so R is quadratic in u. Thus, the coarse-grid correction process is entirely analogous to that for our linear prototype, (1.1). Also, there is no special difficulty here with choice of the subspaces if Ω^{2h} is a conforming subgrid of Ω^h (i.e., n is even) because the corresponding space of functions that are piecewise constant in μ and continuous piecewise linear in x is a subspace of S^h. Note that coarse-grid realizability and recursiveness are natural consequences of this observation.

These ideas were developed in collaboration with Thomas A. Manteuffel. Theory has been developed in this simplified setting to show that $R(u)$ exhibits a sense of ellipticity (i.e., $\|Lu\| \geq \gamma\|u\|$ for $\gamma > 0$ and all suitable u), which in combination with appropriate approximation properties yields discretization error estimates. Also, based on earlier work with other multigrid techniques [Manteuffel, McCormick, Morel, and Yang 1990], initial experiments are currently being conducted by Klaus Ressel using the above relaxation schemes and coarsening strategies. An important aspect of this study will be the use of composite grids to resolve boundary layers, internal layers, and other local phenomena present in many realistic transport models.

4.3.   Galerkin 1: General eigenvalue problems.

The second prototype problem, (1.2), is posed as one of finding the smallest eigenvalue of a self-adjoint linear operator. From certain perspectives, this is fairly simple to treat, especially since a variational principle can be used. Fortunately, many practical problems are of this type, especially in dynamic structural analysis, in single-group neutron diffusion, and in modeling electron energy states. However, even a slight departure from this special case, like small convection terms in L, prohibits a variational formulation. Significant departure, such as when L is hyperbolic or nonlinear, may lose many of the well-behaved properties exhibited by our model: the eigenvalues can become complex, degenerate, or, in many ways, pathological; and the eigenvectors can become complex or generalized, or lose their subspace and completeness properties. It would be a formidable task to deal with these issues here. Instead, we will very briefly outline a *formal* approach to treating eigenvalue problems that does not depend on a variational principle. We will discuss only the formulation of the general problem and the abstract form of its discretization. We will not discuss other aspects of the multilevel projection method because they can be developed much in the same way that they were for (1.2). Also, we will not consider any of the fundamental questions that this formulation raises, such as which eigenvalue is to be approximated, whether continuation methods are necessary, and what grids and subspaces should be used for discretization. The only purpose of this discussion is to suggest a formal procedure for treating eigenvalue problems that are not of the same class as our prototype. In fact, we will focus on a general linear eigenvalue problem, although it should be clear that an analogous formulation may be applied to many nonlinear operators as well.

Consider the generalized eigenvalue problem

$$(4.15) \qquad\qquad Lu = \lambda M u$$

subject to the constraint

$$(4.16) \qquad\qquad \langle u, Nu \rangle = 1.$$

Here, L, M, and N are linear operators defined on an appropriate Hilbert space, H, of functions equipped with the inner product $\langle \cdot, \cdot \rangle$. Let $H_+ = H \times \Re$. Then the main point of this discussion is that (4.15)–(4.16) may be rewritten as the nonlinear equation

$$(4.17) \qquad\qquad K(u_+) = 0,$$

where $u_+ = \begin{pmatrix} u \\ \lambda \end{pmatrix} \in H_+$ and K is the operator on H_+ defined by

$$K \begin{pmatrix} u \\ \lambda \end{pmatrix} = \begin{pmatrix} Lu - \lambda M u \\ \langle u, Nu \rangle - 1 \end{pmatrix}.$$

The discretization of K is based on a finite-dimensional subspace, S^h, of H. Letting $S_+^h = S^h \times \Re$, then the Galerkin discretization of K is $K^h(u_+^h) \equiv P_{S_+^h} K(P_{S_+^h} u_+^h)$, $u_+^h \in S_+^h$. Note that

$$K^h(u_+^h) = K^h \begin{pmatrix} u^h \\ \lambda^h \end{pmatrix} = \begin{pmatrix} L^h u^h - \lambda^h M^h u^h \\ \langle u^h, N^h u^h \rangle - 1 \end{pmatrix},$$

$u^h \in S^h$ and $\lambda^h \in \Re$, where

$$L^h = P_{S^h} L P_{S^h},$$

$$M^h = P_{S^h} M P_{S^h},$$

and

$$N^h = P_{S^h} N P_{S^h}.$$

Now the coarsening of K^h is given in terms of a subspace, S^{2h}, of S^h and defined on $S^{2h}_+ \equiv S^{2h} \times \Re$ by

$$K^{2h}_{u^h_+}(u^{2h}_+) = P_{S^{2h}} K^h(u^h_+ + u^{2h}_+),$$

$u^{2h}_+ \in S^{2h}_+$. Writing $u^{2h}_+ = \begin{pmatrix} u^{2h} \\ \lambda^{2h} \end{pmatrix}$, then

$$K^{2h}_{\begin{pmatrix} u^h \\ \lambda^h \end{pmatrix}} \begin{pmatrix} u^{2h} \\ \lambda^{2h} \end{pmatrix} =$$

$$\begin{pmatrix} L^{2h}u^{2h} - \lambda^{2h}M^{2h}u^{2h} - \lambda^h M^{2h}u^{2h} - \lambda^{2h}P_{S^{2h}}M^h u^h + P_{S^{2h}}(L^h u^h - \lambda^h M^h u^h) \\ \langle u^{2h}, N^{2h}u^{2h}\rangle + \langle u^{2h}, P_{S^{2h}}N^h u^h\rangle + \langle N^h P_{S^{2h}}u^{2h}, u^h\rangle + \langle u^h, N^h u^h\rangle - 1 \end{pmatrix}.$$

Thus, the coarse-grid equation represents a yet more general eigenproblem, but in this generality this projection scheme is evidently coarse-grid realizable and recursive. In fact, these relations are just the Galerkin analogues of the Rayleigh-Ritz scheme applied to the prototype problem, (1.2).

4.4.   Galerkin 2: Riccati equations.

Consider the model *linear quadratic regulator* problem, which is defined in terms of the *state*, $v = v(x,t)$, and *control*, $w = w(t)$, by

Minimize

$$\int_0^\infty \int_0^1 (vQv + rw^2)(x,t)\, dx\, dt$$

subject to

$$v_t = Av + bw \quad \text{on } [0,1] \times [0,\infty),$$
$$v(0,t) = v(1,t) = 0, \quad t > 0,$$
$$v(x,0) = v_0(x), \quad 0 \le x \le 1.$$

Here, Q is assumed to be an integral operator of the form

$$Qz(x) = \int_0^1 q(x,\zeta)z(\zeta)\, d\zeta;$$

r is a constant, so without loss of generality we take $r = 1$; A is the one-dimensional operator $\nabla \cdot \nabla = \frac{\partial^2}{\partial x^2}$; and $b = b(x) \geq 0$ and $v_0(x)$ are given. The spatial functions are assumed to lie in some Hilbert space, H (e.g., $H = L_2[0,1]$). The control solution, w, of this problem may be written formally in terms of the linear operator, U, applied to the state:

$$w = B^*Uv.$$

Here we define the operator $B : \Re \to H$ by

$$B\alpha = \alpha b,$$

$\alpha \in \Re$, so that $B^* : H \to \Re$ is the integral operator defined by

$$B^*w = \int_0^1 b(\zeta)w(\zeta)\,d\zeta.$$

The task is to find U, which may be a solution of the *Riccati equation* (assuming U and Q are self-adjoint):

$$AU + UA - UBB^*U + Q = 0.$$

Now let L be the two-dimensional Laplacian, $L = -\nabla \cdot \nabla = -\Delta$. Suppose that U can be written as an integral operator, given by

$$Uz(x) = \int_0^1 u(x,\eta)z(\eta)\,d\eta,$$

$z \in H$. Suppose also that q and u are symmetric kernels in the sense that

$$q(x,y) = q(y,x)$$

and

$$u(x,y) = u(y,x),$$

$0 < x,\ y < 1$. Suppose finally that q is nonnegative and u is positive in Ω. Then substitution of the differential expression for A and the integral representations for U, B^*, and Q into the Riccati equation leads to the following equation, whose positive symmetric solution is presumably the kernel of U:

$$(4.18) \qquad K(u) \equiv Lu + (B_xu)(B_yu) - q = 0, \quad u \in H,$$

where

$$B_xu(x,y) = \int_0^1 b(\zeta)u(\zeta,y)\,d\zeta$$

and

$$B_y u(x, y) = \int_0^1 b(\eta) u(x, \eta) \, d\eta.$$

Our discretization of K uses a uniform grid, Ω^h, on Ω and the subspace, $S^h \subset H$, of continuous piecewise bilinear functions with respect to its cells. The discrete operator is defined by $K^h(u^h) = P_{S^h} K(P_{S^h} u^h)$, $u^h \in S^h$. Note that

(4.19) $\qquad K^h(u^h) = L^h u^h + P_{S^h}(B_x P_{S^h} u^h)(B_y P_{S^h} u^h) - q^h,$

where $L^h = P_{S^h} L P_{S^h}$ corresponds to the matrix defined in (1.28) and $q^h = P_{S^h} q$ is the usual finite element source-term discretization (cf. (1.29)). To examine the nonlinear term in (4.19), let $w^h_{(ij)}$, $1 \le i,\, j \le n$, denote the hat functions in S^h defined in Chapter 1, §1.5. We can write

$$w^h_{(ij)}(x, y) = w^h_{(i)}(x) w^h_{(j)}(y),$$

where $w^h_{(i)}$ is the one-dimensional hat function associated with the uniform grid with mesh size h defined on $[0, 1]$, $1 \le i,\, j \le n$. Define

(4.20) $\qquad b_i^h = \int_0^1 b(\zeta) w^h_{(i)}(\zeta) \, d\zeta,$

$1 \le i \le n$. These coefficients will usually require approximation, unless the function b has special properties, for example, when b is a piecewise polynomial with respect to the cells of grid h on $[0, 1]$. Note, for piecewise constant b, that $b_i^h = (b((i-1/2)h) + b((i+1/2)h))h/2$, $1 \le i \le n$. Then, writing

$$u^h = \sum_{p,q=1}^n u^h_{pq} w^h_{(pq)}$$

and fixing $i, j \in \{1, 2, \cdots, n\}$, we have

$$\langle w^h_{(ij)}, (B_x u^h)(B_y u^h) \rangle$$

$$= \left\langle w^h_{(ij)}, \left(B_x \sum_{p,q=1}^n u^h_{pq} w^h_{(pq)} \right) \left(B_y \sum_{p,q=1}^n u^h_{pq} w^h_{(pq)} \right) \right\rangle$$

$$= \int_0^1 \int_0^1 w^h_{(i)}(x) w^h_{(j)}(y) \left[\sum_{p,q=1}^n u^h_{pq} b^h_p w^h_{(q)}(y) \right] \left[\sum_{p,q=1}^n u^h_{pq} b^h_q w^h_{(p)}(x) \right]$$

$$= \left[\sum_{p,q=1}^n b^h_p \langle w^h_{(j)}, w^h_{(q)} \rangle u^h_{pq} \right] \left[\sum_{p,q=1}^n b^h_q \langle w^h_{(i)}, w^h_{(p)} \rangle u^h_{pq} \right].$$

Here, the one-dimensional inner products satisfy

$$\langle w^h_{(k)}, w^h_{(\ell)} \rangle = \begin{cases} 2h/3 & k = \ell \\ h/6 & |k - \ell| = 1 \\ 0 & \text{otherwise.} \end{cases}$$

Hence, the nodal representation for $P_{S^h}(B_x P_{S^h} u^h)(B_y P_{S^h} u^h)$ is given by

$$(\underline{B_x^h u^h}) * (\underline{B_y^h u^h}),$$

where, for $1 \leq i, j \leq n$: $*$ denotes vector products computed componentwise (i.e., $(\underline{v^h} * \underline{w^h})_{(ij)} = v_{(ij)}^h w_{(ij)}^h$); $\underline{u^h}$ is the nodal representation of u^h; $\underline{B_x^h}$ is the matrix defined by

$$(4.21) \qquad (\underline{B_x^h u^h})_{(ij)} = \frac{h}{6} \sum_{p=1}^{n} b_p^h (u_{pj-1}^h + 4u_{pj}^h + u_{pj+1}^h);$$

and $\underline{B_y^h}$ is defined analogously by

$$(4.22) \qquad (\underline{B_y^h u^h})_{(ij)} = \frac{h}{6} \sum_{q=1}^{n} b_q^h (u_{i-1q}^h + 4u_{iq}^h + u_{i+1q}^h).$$

The discrete equation can thus be written as

$$(4.23) \qquad \underline{K^h}(\underline{u^h}) \equiv \underline{L^h u^h} + (\underline{B_x^h u^h}) * (\underline{B_y^h u^h}) - \underline{q^h} = 0.$$

Because of the quadratic nature of the nonlinearity, implementing point Gauss-Seidel on (4.23) is fairly straightforward. In analogy to (2.18), the ijth step involves finding an $s \in \Re$ that solves

$$(4.24) \qquad \langle \underline{w_{(ij)}^h}, \underline{K^h}(\underline{u^h} + s\underline{w_{(ij)}^h}) \rangle = 0.$$

It is easy to see that

$$\langle \underline{w_{(ij)}^h}, \underline{K^h}(\underline{u^h} + s\underline{w_{(ij)}^h}) \rangle = r_{ij}^h + d_{ij}^h s + c_{ij}^h s^2,$$

where

$$(4.25) \qquad \begin{cases} r_{ij}^h = \langle \underline{w_{(ij)}^h}, \underline{K^h}(\underline{u^h}) \rangle = (\underline{L^h u^h} + (B_x u^h) * (B_y u^h) - \underline{q^h})_{(ij)}, \\ d_{ij}^h = L_{ij-ij}^h + \frac{2h}{3} b_j^h (\underline{B_x^h u^h})_i + \frac{2h}{3} b_i^h (\underline{B_y^h u^h})_j, \\ c_{ij}^h = \frac{4h^2}{9} b_i^h b_j^h. \end{cases}$$

Hence, one root of (4.24) is given (in numerically accurate form) by

$$(4.26) \qquad s = -\frac{2r_{ij}^h}{d_{ij}^h + \sqrt{(d_{ij}^h)^2 - 4r_{ij}^h c_{ij}^h}}.$$

This root corresponds to the correct step size in the linear limiting case $b \equiv 0$.

The computational complexity of the discretization in (4.23), which is the cost of evaluating $\underline{K^h}(\underline{u^h})$ for a typical $\underline{u^h}$, is $O(n^2)$. This follows from observing

that the nonlinear terms in $\underline{K}^h$ have the same $O(n^2)$ complexity as the linear
terms: for example, the quantity in (4.21) costs $O(n)$ to compute, but it needs
to be evaluated only for each of the n rows of u (i.e., for $j = 1, 2, \cdots, n$). In this
light, a straightforward implementation of point Gauss-Seidel is generally too
expensive because the quantities in (4.25) must be computed at a cost of $O(n)$
operations for each of the n^2 grid points, yielding a total cost per sweep of $O(n^3)$.
The culprit here is that the terms in (4.21) and (4.22) must be recomputed for
each of the n^2 points. When the nonlinear terms in (4.23) do not dominate
the discretization, a less costly scheme is determined by the linear limiting case,
which gives the step size

$$s = -\frac{r_{ij}^h}{L_{ij-ij}^h}.$$

It is, however, generally more effective to use (4.26) but freeze the nonlinear
terms computed in (4.25) over each full relaxation sweep. Another possibility is
to use a global relaxation scheme akin to Richardson's or Jacobi's method. For
example, $\underline{u}^h$ could be corrected in the direction $\underline{w}^h \equiv \underline{K}^h(u^h)$ using the step-size
criterion

$$\langle \underline{w}^h, \underline{K}^h(\underline{u}^h + s\underline{w}^h) \rangle = 0.$$

A third possibility is simply to implement point Gauss-Seidel by computing the
quantities in (4.25) initially for all points, then using inexpensive update formulas
to correct them locally after the iteration step at each point is performed. In
any case, all of these alternatives are $O(n^2)$ in complexity and should exhibit
good smoothing, provided the nonlinear terms in the discretization are not too
dominant. (Point relaxation is generally a poor smoother with respect to the
nonlinear terms, so other relaxation processes must be considered when these
terms dominate.)

Assume that $n > 1$ is odd, so that the uniform subgrid, $\Omega^{2h} \subset \Omega^h$, conforms
to Ω. Let S^{2h} be the subspace of S^h corresponding to Ω^{2h}. Then the coarse-grid
operator is defined in principle by

$$K_{u^h}^{2h}(u^{2h}) \equiv P_{S^{2h}} K^h(u^h + u^{2h}),$$

$u^{2h} \in S^{2h}$. Let $r^h = K^h(u^h)$ and $q^{2h} = P_{S^{2h}} r^h$. Define B_x^{2h} and B_y^{2h} in analogy
to the respective definitions of B_x^h and B_y^h in (4.21) and (4.22). (Note that if
b is piecewise constant with respect to the cells of Ω^h, then by (4.20) we have
$b_{2i}^{2h} = (b((2i - \frac{3}{2})h) + 3b((2i - \frac{1}{2})h) + 3b((2i + \frac{1}{2})h) + b((2i + \frac{3}{2})h))\frac{h}{8}$.) Now define
the functions $\beta_x^{2h} \equiv P_{S^{2h}} B_x^h u^h$ and $\beta_y^{2h} \equiv P_{S^{2h}} B_y^h u^h$ and the grid $2h$ linear
operator

$$M_{u^h}^{2h} \equiv L^{2h} + \beta_x^{2h} B_y^{2h} + \beta_y^{2h} B_x^{2h}.$$

Then it is easy to verify that

$$K_{u^h}^{2h}(u^{2h}) = M_{u^h}^{2h} u^{2h} + (B_x^{2h} u^{2h})(B_y^{2h} u^{2h}) - q^{2h}.$$

Thus, the coarse-grid operator has a form (and complexity!) analogous to the fine-grid operator, except that the linear part is modified by linear integral operators. In any event, this establishes both coarse-grid realizability and recursiveness.

The prospects for developing an effective Riccati solver based on multilevel projection methods seem to be very good. If the usual multilevel convergence rates can be achieved, then a total computational complexity of order n^2 should generally be attainable. That is, we should be able to compute the matrix corresponding to the Riccati integral operator to an accuracy comparable to discretization error at a cost proportional to the number of nonzero coefficients in the fine-grid operator. This is the usual complexity property of multigrid methods. Actually, we may even be able to do much better with local refinement. To suggest this, it may be in many cases that the Riccati kernel $u(x,y)$ is smooth as a function of x and y when $|x - y| \gg 0$. For such x and y, u can therefore be well approximated on coarser levels, and it should be possible to use a composite grid that is increasingly finer near the diagonal line $x - y = 0$. Figure 4.3 depicts an example based on using triangular elements, which conform better than rectangles to the diagonal geometry. In cases where enough accuracy can be obtained by restricting the finer levels to just a few elements on either side of the diagonal, we may be able to develop a PML scheme for computing the Riccati kernel with a total computational complexity on the order of $n \log n$. That is, by using a composite grid with $O(n \log n)$ points, we may be able to produce an approximation with actual accuracy comparable to that of the solution of the $n \times n$ global fine grid.

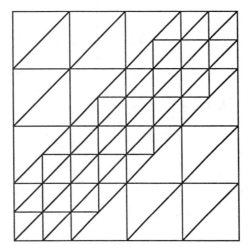

FIG. 4.3. *Example two-level composite grid for the Riccati kernel.*

The research work on this topic is in its early stages, with only preliminary numerical results obtained thus far (cf. [Ito, McCormick, and Tiejun 1991]). Encouraged by early success for the simpler cases, the ultimate aim is to see how far these principles will carry into higher dimensions (e.g., four- and six-dimensional equations of the form (4.18)) and more sophisticated PDEs (e.g., advection-diffusion and Navier-Stokes equations).

4.5.   Petrov-Galerkin 1: The finite volume element method (FVE).

In this section, we develop a multilevel projection method based on the finite volume element (FVE) discretization technique, which was used in the Frontiers book [McCormick 1989] to treat fluid flow equations. We use the interpretation of FVE as a Petrov-Galerkin scheme to provide the basis for GML. The development may seem a little awkward because the natural approach to coarsening (i.e., the natural choice of coarse-grid spaces) leads to technical difficulties (e.g., nonconforming elements)—and this we avoid by taking a coarsening approach that is perhaps more algebraic than physical. The method that results from this approach is different from, and less natural than, the physically based multigrid scheme developed in the Frontiers book. Nevertheless, this paradigm is useful if for no other reason than to illustrate one possible general mechanism for dealing with nonconforming elements.

Consider the two-dimensional Poisson equation

$$(4.27) \qquad\qquad K(u) \equiv -\Delta u - f = 0, \quad u \in H,$$

where H is an appropriate Hilbert space of functions defined on the unit square, $\Omega \subset \Re^2$, that have zero values on $\partial \Omega$. Actually, we take (4.27) in the following weak sense. For any suitable control volume, $V \subset \Omega$, define its *characteristic function* by

$$v_V(z) = \begin{cases} 1 & z \in V \\ 0 & \text{otherwise.} \end{cases}$$

Note formally that

$$\begin{aligned}
\langle v_V, K(u) \rangle &= \int_\Omega v_V(z)(-\Delta u(z) - f(z))\, dz \\
&= \int_V (-\nabla \cdot \nabla u(z) - f(z))\, dz \\
&= \int_{\partial V} \underline{n} \cdot (-\nabla u(s))\, ds - \int_V f(z)\, dz,
\end{aligned}$$

where $\underline{n}$ is the outward unit normal on ∂V. Define

$$k(v_V, u) = \int_{\partial V} n \cdot (-\nabla u(s))\, ds - \int_V f(z)\, dz.$$

Then by (4.27) we really mean the weak form

$$(4.28) \qquad\qquad k(v_V, u) = 0, \quad u \in H_1, \quad \forall v_V \in H_2,$$

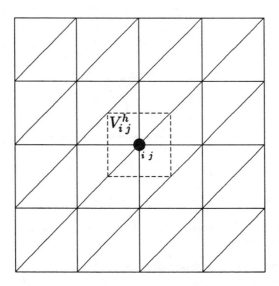

FIG. 4.4. *Grid, triangulation, and sample control volume.*

where H_1 is possibly a different Hilbert space of functions defined on Ω with zero values on $\partial\Omega$, and H_2 is the set of characteristic functions of suitable control volumes in Ω.

The discretization of (4.28) is based on: a uniform grid, $\Omega^h = \{z_{ij} : 1 \leq i, j \leq n\}$; the space of continuous piecewise linear functions, $S^h \subset H_1$, associated with a triangulation of Ω^h; and the set of characteristic functions of the square control volumes corresponding to Ω^h, which we write as $T^h = \{v^h_{(ij)} : 1 \leq i, j \leq n\}$ (i.e., $v^h_{(ij)} = v_{V^h_{ij}}$). See Figure 4.4. Then the discrete form is defined by

$$k^h(v^h_{(ij)}, u^h) \equiv k(v^h_{(ij)}, P_{S^h} u^h),$$

$u^h \in S^h$, $1 \leq i, j \leq n$. The discretization of (4.27) is thus given by

(4.29) $$k^h(v^h_{(ij)}, u^h) = 0, \quad u^h \in S^h, \quad \forall i, j \in \{1, 2, \cdots, n\}.$$

It is easy to see that this can be expressed in terms of the node values of u^h as

(4.30) $$-u^h_{i+1\,j} - u^h_{i\,j+1} + 4u^h_{i\,j} - u^h_{i-1\,j} - u^h_{i\,j-1} = h^2 f^h_{i\,j},$$

where we write $f^h_{ij} = \frac{1}{h^2} \int_{V^h_{ij}} f(z)\,dz$ to suggest that $f^h_{ij} \sim f(z_{ij})$. This is of course the usual five-point difference formula for discretizing (4.27).

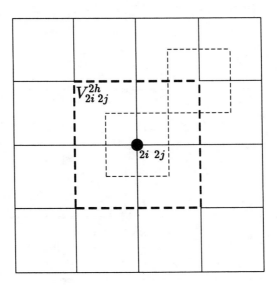

FIG. 4.5. *Physically defined coarse-grid volume (bold dashes) and related fine-grid volumes (regular dashes).*

The natural choice for relaxation on (4.29) is point Gauss-Seidel, which is known to be an effective smoother in this context. Unfortunately, the natural choice for coarsening is problematic. To see this, suppose that $n > 1$ is odd, so that the uniform subgrid Ω^{2h} and its triangulation are conforming. The problem here is that the *physical* volumes that we would naturally choose with Ω^{2h} do not conform to V^h. See Figure 4.5 (note the use of grid h indices to label grid $2h$), which shows that the physical volume $V^{2h}_{2i\,2j}$ cannot be written as a union of the $V^h_{k\,\ell}$ (i.e., the characteristic function $v^{2h}_{(2i\,2j)}$ defined in this physical way is not in the space T^h). This does not formally prevent us from applying the multilevel projection method. In fact, we could define the coarse-grid form by

$$k^{2h}_{u^{2h}}(v^{2h}_{(2i\,2j)}, u^{2h}) \equiv k(v^{2h}_{(2i\,2j)}, u^h + u^{2h}),$$

$u^{2h} \in S^{2h}$, $1 \le i, j \le (n-1)/2$. Note that k must be used here because $k^h(v^{2h}_{(2i\,2j)}, u^h + u^{2h})$ is not defined, even formally. Unfortunately, this definition introduces two serious difficulties, the first of which is that the expression $k(v^{2h}_{(2i\,2j)}, u^h + u^{2h})$ involves partial derivatives of u^h at element boundaries. We could simply define these derivatives as, say, the average of left-sided and right-sided derivatives there, but we would then face the more serious difficulty that $k^{2h}_{u^h}$ is not generally consistent with k^h: by definition, the exact solution of (4.29) satisfies the conservation law (4.28) on grid h volumes, but we cannot expect it to satisfy (4.28) on grid $2h$ volumes. This means that the coarse-grid correction would tend to divert the fine-grid process from converging to the exact (fine-grid) solution.

A remedy for these difficulties is to abandon the physical coarsening process in favor of a conforming algebraic one. This is done simply as follows. Define the *algebraic* coarse-grid functions

$$v^{2h}_{(2i\ 2j)} = \frac{1}{4}(v^h_{(2i+1\ 2j+1)} + v^h_{(2i-1\ 2j+1)} + v^h_{(2i+1\ 2j-1)} + v^h_{(2i-1\ 2j-1)})$$
$$+ \frac{1}{2}(v^h_{(2i+1\ 2j)} + v^h_{(2i-1\ 2j)} + v^h_{(2i\ 2j+1)} + v^h_{(2i\ 2j-1)}) + v^h_{(2i\ 2j)},$$

$1 \le i, j \le (n-1)/2$, which we can think of as conforming approximations to the physical characteristic functions. We then simply define the coarse-grid form by

$$k^{2h}_{u^h}(v^{2h}_{(2i\ 2j)}, u^{2h}) \equiv k^h(v^{2h}_{(2i\ 2j)}, u^h + u^{2h}),$$

$1 \le i, j \le (n-1)/2$. This eliminates the trouble with conformity because, by construction, each $v^{2h}_{(2i\ 2j)}$ is in the span of T^h. We now show that $k^{2h}_{u^h}$ is realizable on grid $2h$ in a recursive way.

We do this in two stages. In the first, a simple calculation shows that the coarse-grid equation

$$k^{2h}_{u^h}(v^{2h}_{(2i\ 2j)}, u^h + u^{2h}) = 0, \quad u^{2h} \in S^{2h}, \quad \forall i,j \in \{1,2,\cdots,(n-1)/2\},$$

can be written in terms of the *grid h node values* as the stencil equation

$$\begin{bmatrix} 0 & -\frac{1}{4} & -\frac{1}{2} & -\frac{1}{4} & 0 \\ -\frac{1}{4} & 0 & \frac{1}{2} & 0 & -\frac{1}{4} \\ -\frac{1}{2} & \frac{1}{2} & 2 & \frac{1}{2} & -\frac{1}{2} \\ -\frac{1}{4} & 0 & \frac{1}{2} & 0 & -\frac{1}{4} \\ 0 & -\frac{1}{4} & -\frac{1}{2} & -\frac{1}{4} & 0 \end{bmatrix} u^{2h}_{(2i\ 2j)} = - \begin{bmatrix} 0 & 0 & 0 & 0 & 0 \\ 0 & \frac{1}{4} & \frac{1}{2} & \frac{1}{4} & 0 \\ 0 & \frac{1}{2} & 1 & \frac{1}{2} & 0 \\ 0 & \frac{1}{4} & \frac{1}{2} & \frac{1}{4} & 0 \\ 0 & 0 & 0 & 0 & 0 \end{bmatrix} r^h_{(2i\ 2j)},$$

$1 \le i, j \le (n-1)/2$, where we use the residuals for (4.30) defined by

$$r^h_{(p\ q)} = (-u^h_{p+1\ q} - u^h_{p\ q+1} + 4u^h_{p\ q} - u^h_{p-1\ q} - u^h_{p\ q-1}) - h^2 f^h_{p\ q},$$

$1 \le p, q \le n$. In the second stage, we eliminate all but grid $2h$ node values on the left-hand side of this expression by using the fact that $u^{2h} \in S^{2h}$ (e.g., $u^{2h}_{(2i+1\ 2j)} = \frac{1}{2}(u^h_{(2i\ 2j)} + u^h_{(2i+2\ 2j)}))$. This yields the following matrix equation written in terms of the *grid $2h$ node values*:

$$\underline{L}^{2h}\underline{u}^{2h} = \underline{f}^{2h},$$

where (using grid $2h$ indices)

$$(L^{2h})_{ij-pq} = \begin{cases} 3 & |i-p| = |j-q| = 0 \\ -\frac{1}{2} & |i-p| + |j-q| = 1 \\ -\frac{1}{4} & |i-p| = |j-q| = 1 \\ 0 & \text{otherwise} \end{cases}$$

and

$$\underline{f}^{2h} = -\underline{L}_h^{2h}\underline{r}^h,$$

with $\underline{I}_h^{2h}$ the scaled version of the usual full-weighting operator:

$$(\underline{I}_h^{2h})_{ij-pq} = \begin{cases} 1 & |i-p| = |j-q| = 0 \\ \frac{1}{2} & |i-p| + |j-q| = 1 \\ \frac{1}{4} & |i-p| = |j-q| = 1 \\ 0 & \text{otherwise.} \end{cases}$$

Clearly, then, $k_{u^h}^{2h}$ is grid $2h$ realizable. That this can be done recursively follows from noting that the two-stage process can be repeated, but now applied to the nine-point matrix $\underline{L}^{2h}$ instead of $\underline{L}^h$.

The coarse-grid equations for this scheme can be dynamically generated virtually for any weak form k by computationally averaging the equations (first stage), then reducing the unknowns (second stage). In many cases, this can be done analytically as we did here for the two-level case. However, this approach seems somewhat less tractable than the physically based multigrid schemes developed for FVE in the Frontiers book. Nevertheless, this example still serves to illustrate a general rule for treating coarsening processes that are not naturally conforming. The basic idea is to approximate the nonconforming elements by a subspace of the fine-grid space. Many existing methods for nonconforming coarse-grid correction can be interpreted in this framework.

4.6.   Petrov-Galerkin 2: Image reconstruction.

The ideas developed here originated in the early 1980s, although only a primitive form of the multigrid scheme was ever implemented (cf. [Herman, Levkowitz, McCormick, and Tuy 1982]). Actually, it is unclear how much advantage there is in a multilevel approach since transform-based methods are very efficient, at least for many conventional applications. Also, without implementing it, we cannot begin to predict how effective the multilevel scheme might be since we are very much on uncharted ground. (See, however, the preliminary but encouraging results for a related approach described in [Hudson 1990].) Nevertheless, in addition to its intrinsic importance, the image reconstruction paradigm is instructive because it exposes the real and separate roles of coarsening the equations and the unknowns.

As a simplified model, consider a two-dimensional square *object*, which is assumed to be represented by a nonnegative *density* function, u, in $H_1 \equiv L_2(\Omega)$, where Ω is the unit square in $\Re^2$. Suppose that a *projection* of the object is taken by forming line integrals, or *rays*, of the function along all parallel paths through the object at a given angle to the x-axis. See Figure 4.6, which depicts one of these rays for one projection taken at a $45°$ angle. Note that this specific projection is represented by the expression

(4.31) $$\sqrt{2} \int_{\max(0,z)}^{\min(1,1+z)} u(s, s-z)\, ds = f_{(45°)}(z),$$

$z \in [-1, 1]$. Thus, the object, u, gives rise to projections, $f_{(\alpha)}$, taken as the angle, α, varies over some fixed set of finite cardinality, $p \geq 1$. This yields the linear operator equation

$$(4.32) \qquad\qquad Lu = f, \quad u \in H_1,$$

where f is a block function consisting of the $f_{(\alpha)}$ and $L : H_1 \to H_2$ is the operator of corresponding line integrals of the form (4.31). Here, H_2 is the block space of one-dimensional L_2 functions corresponding to the line integrals $f_{(\alpha)}$ defined on their respective projection intervals. The task now is to determine, or *reconstruct*, u in (4.32) given its projections, f, with the added constraint

$$(4.33) \qquad\qquad u \geq 0.$$

Note that a solution of (4.32) exists only if f satisfies the *compatibility condition*

$$(4.34) \qquad\qquad \int f_{(\alpha)}(z)\, dz \equiv c,$$

where c is a constant independent of α. (We avoid specifying the actual limits of integration because they depend on α, and we prefer not to introduce additional notational complexity.) This condition must be met because the integral of the left-hand side of (4.31) over z is independent of α. We assume henceforth that (4.34) is satisfied, which is necessary for f to be in the range of L.

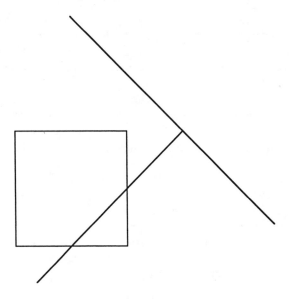

FIG. 4.6. *Object, projection at $45°$, and one of its rays.*

Mathematical models of the form in (4.32) arise in x-ray tomography, for example. The object function, u, thus represents the density of an unknown *image*, perhaps a cross-section of a human head, and each projection is obtained by one x-ray exposure. Note that (4.34) is then interpreted physically as the requirement that each x-ray exhibit the same average *gray* level, which should be true in an idealized experiment. In any event, one of the major difficulties in solving (4.32) is that such problems are highly underdetermined, which stems from the fact that the goal is to reconstruct a two-dimensional image from a finite set of one-dimensional projections. Thus, when (4.32) has a solution, it must have infinitely many. One of the difficult numerical tasks is therefore to identify and compute the solution that best represents the physical process. It is not our goal here to enter long discussions on issues specific to the application, so instead we will just assume that the solution we seek is the one closest (in L_2 sense) to the uniformly gray image $u \equiv c$, where c is the constant given in (4.34). Thus we are looking for the *minimal-norm* solution of (4.32), which is the unique solution orthogonal to the null space of L. (This follows from the fact that $u \equiv c$ is orthogonal to the null space of L, which can be deduced from the easily verified relationship $L^*Lc = pc$.)

The general nature of (4.32) suggests that the discretizations are likely to give rise to rectangular matrices. More precisely, the discrete problem will not in general exhibit the usual one-to-one correspondence between equations and unknowns that is typical of PDE applications. In this sense, the image reconstruction problem represents a major departure from the other paradigms of this chapter; it will lead us to relaxation schemes other than Gauss-Seidel and to discretization and coarsening strategies that treat the equations and unknowns separately.

Another major difficulty with reconstruction problems is the typically large size of their discretizations. Usually, real applications involve three dimensions with resolution requirements in the millions or billions of image cells, or *pixels*. The matrices therefore tend to have millions or billions of rows and columns. Moreover, although they are very sparse, there is generally very little structure that can be used to advantage by anything other than simple transform-based or relaxation-type methods. It is with this computational scale in mind that we now describe one possible multilevel method for solving (4.32).

Discretization of (4.32) depends on two subspaces, S^h and T^h. We discretize the image and projection spaces by forming uniform arrays of pixels, on which we assume that the corresponding functions are constant. This means that S^h and T^h consist of piecewise constant functions corresponding to the pixels, or cells, of uniform grids placed on Ω and the projection intervals, respectively. Figure 4.7 shows these grids for the image and one projection. Now let the image grid be of dimension $n \times n$ and let the projection grid be of dimension $m \times 1$, where $n \geq 1$ and $m \geq p \geq 1$. Then the discrete ray depicted in Figure 4.7 corresponds to one of the equations in the discrete system

$$(4.35) \qquad\qquad L^h u^h = f^h, \quad u^h \in S^h,$$

where $L^h = P_{T^h} L P_{S^h}$ and $f^h = P_{T^h} f$. Note that the row corresponding to this discrete ray has nonzero entries only in columns corresponding to cells that the ray intersects, with entries equal to the area of intersection. We emphasize here that the number, p, of projections is unchanged by the discretization. Note also that our notation suggests that the same mesh size is used for the image and projection grids. This seems natural, but it may be that different mesh sizes are more effective for characterization of the physical solution of (4.32). In fact, it is likely that the relationship between the two mesh sizes is important to the well-posedness of the discretization. Note, for example, that (4.35) becomes increasingly either singular or overdetermined, depending on whether the image-to-projection mesh size ratio is either increased or decreased.

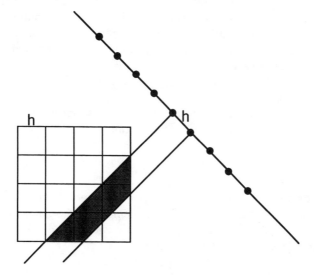

FIG. 4.7. *Discretization grids for the image and one projection, including one of its discrete rays.*

For relaxation we choose Kaczmarz's method (see Chapter 2, §2.5), which is commonly used in iterative reconstruction algorithms:

For $i = 1, 2, \cdots, m$:

Solve

(4.36) $$\langle w_{(i)}^h, L^h(u^h + s^h(L^h)^* w_{(i)}^h) - f^h \rangle = 0$$

and set $u^h \leftarrow u^h + s^h(L^h)^* w_{(i)}^h$.

Here, $w_{(i)}^h$ is the characteristic function of the ith projection pixel (i.e., one of the cells of one of the projections), $1 \le i \le m$, and $(L^h)^*$ is the adjoint of $L^h : S^h \to T^h$. Note that the step size that satisfies (4.36) is given by

$$s^h = -\frac{\langle w_{(i)}^h, L^h u^h - f^h \rangle}{\|(L^h)^* w_{(i)}^h\|^2},$$

$1 \leq i \leq n$.

Kaczmarz has a simple physical interpretation: the ith step involves projecting the current image approximation onto the ith pixel using the ith ray and determining the deviation of the result from the required gray level, then *back-projecting* a correction by subtracting this deviation from each image pixel value in proportion to the ratio of the areas of intersection of the ray with the given pixel and with the whole image. In other words, the value of each image pixel is changed by a constant times the area of ray intersection, where the constant is chosen to make the ith ray yield the correct projection.

Coarsening is not as straightforward as discretization and relaxation. We are used to developing multigrid schemes that have a natural correspondence between individual, or blocks of, equations and unknowns. For image reconstruction problems, no such correspondence exists, even approximately. For relaxation, this made us abandon Gauss-Seidel in favor of a method that allows for more general matrices, for which Kaczmarz is more or less an obvious choice. But for coarsening, this lack of correspondence forces us into a fundamental reconsideration of the different roles of S^h and T^h. This in turn leads us to examine relaxation more closely.

To this end, let $\underline{A}$ be a real $m \times n$ matrix and suppose that the m-vector $\underline{b}$ is in its range. Then the $m \times m$ matrix equation

$$(4.37) \qquad \underline{A}\,\underline{x} = \underline{b}, \quad \underline{x} \in \Re^h,$$

is solvable. Remember that we are seeking minimal-norm solutions. But such a solution of (4.37) is in the orthogonal complement of the null space of $\underline{A}$, so it is in the range of $\underline{A}^t$. Then solving (4.37) is equivalent to determining $\underline{x} = \underline{A}^t\underline{y}$, where $\underline{y}$ solves the $m \times m$ matrix equation

$$(4.38) \qquad \underline{A}\,\underline{A}^t\underline{y} = \underline{b}, \quad \underline{y} \in \Re^m.$$

Now since $\underline{A}\,\underline{A}^t$ is symmetric and nonnegative definite, then we can at least formally apply Gauss-Seidel to (4.38), with the ith step given by

Solve

$$(4.39) \qquad \langle \underline{w}_{(i)}, \underline{A}\,\underline{A}^t(\underline{y} + s\underline{w}_{(i)}) - \underline{b}\rangle = 0$$

and set $\underline{y} \leftarrow \underline{y} + s\underline{w}_{(i)}$.

Here, $\underline{w}_{(i)}$ is the ith coordinate vector in $\Re^m$. The relationship $\underline{x} = \underline{A}^t\underline{y}$ then shows the equivalence of (4.36) and (4.39).

The main point of this discussion is that Kaczmarz relaxation applied to (4.35) is essentially equivalent to Gauss-Seidel applied to

$$(4.40) \qquad L^h(L^h)^*v^h = f^h, \quad v^h \in T^h.$$

One implication of this equivalence is that (4.40) can be used to guide the coarsening of (4.35). But (4.40) can be formally coarsened according to the multilevel projection method, which yields the correction equation (for fixed $v^h \in T^h$)

$$(4.41) \qquad P_{T^{2h}} L^h (L^h)^* P_{T^{2h}} v^{2h} = f^{2h} \equiv P_{T^{2h}} (f^h - L^h v^h), \quad v^{2h} \in T^{2h}.$$

Let $w^h = (L^h)^* P_{T^{2h}} v^{2h}$ and define

$$(4.42) \qquad L^{2h} = P_{T^{2h}} L^h.$$

Then (4.41) corresponds to the correction equation for (4.35) given by

$$(4.43) \qquad L^{2h} w^h = f^{2h}, \quad w^h \in S^h.$$

This equation represents a legitimate coarsening for (4.35). Remember that we are looking for a minimal-norm solution, which means that the correct w^h will be in the range of $(L^{2h})^*$. This can be computed by starting from the initial guess $w^h = 0$ and using Kaczmarz's method (which can in turn be accelerated by coarse-level corrections based on T^{4h}). Once this w^h has been computed, then the new approximation on level h is given by the replacement process

$$(4.44) \qquad u^h \leftarrow u^h + w^h.$$

The interesting aspect of this correction is that it involves coarsening the equations only—not the unknowns. This may seem unusual, but its basic soundness is substantiated by the development above, which exploited the equivalence between (4.35) and (4.40).

It is interesting to carry these thoughts a little further. Suppose for the moment that we implement the PML process on (4.35) by performing one Kaczmarz sweep on level h followed by one sweep on level $2h$. The coarse-level step is made by choosing $w^h = 0$, performing one Kaczmarz sweep on (4.43), then making the correction in (4.44). It is then easy to see that the complete PML process is equivalent to performing one Kaczmarz sweep on the new block system

$$\begin{pmatrix} L^h \\ L^{2h} \end{pmatrix} u^h = \begin{pmatrix} f^h \\ f^{2h} \end{pmatrix}, \quad u^h \in S^h,$$

where $f^{2h} \equiv P_{T^{2h}} f^h$. In fact, a $(1,0)$ V-cycle using the $q+1$ subspaces $T^{2^q h} \subset T^{2^{q-1} h} \subset \cdots \subset T^{2h} \subset T^h$, $q \geq 0$, is equivalent to just one Kaczmarz sweep applied to the system

$$(4.45) \qquad M^h u^h = g^h, \quad u^h \in S^h,$$

where the block quantities are given by

$$M^h = \begin{pmatrix} L^h \\ L^{2h} \\ \vdots \\ L^{2^q h} \end{pmatrix} \equiv \begin{pmatrix} L^h \\ P_{T^{2h}} L^h \\ \vdots \\ P_{T^{2^q h}} L^h \end{pmatrix}$$

and

$$g^h = \begin{pmatrix} f^h \\ f^{2h} \\ \vdots \\ f^{2^q h} \end{pmatrix} \equiv \begin{pmatrix} f^h \\ P_{T^{2h}} f^h \\ \vdots \\ P_{T^{2^q h}} f^h \end{pmatrix}.$$

In other words, to implement a multilevel method for solving (4.35), it is enough simply to add lumped equations to the system and just continue to apply Kaczmarz to the augmented equations.

The role of coarsening the equations is to attempt to accelerate the Kaczmarz relaxation process, but it does little else to reduce computational cost: one Kaczmarz sweep on level $2h$ is about as expensive as it is on level h, as a close inspection of Figure 4.8 suggests. This cost cannot be reduced without using a coarser image space. In fact, reducing the cost per sweep is the real role of coarsening the unknowns. The subspace, S^{2h}, is necessary only if sweeps on (4.43) are considered to be too expensive. In this case we replace (4.43) by the correction equation

$$L^{2h} u^{2h} = f^{2h}, \quad u^{2h} \in S^{2h},$$

where, instead of (4.42), we use the fully coarsened operator $L^{2h} = P_{T^{2h}} L^h P_{S^{2h}}$. (See Figure 4.9.) The expectation here is that u^{2h} is a good enough approximation to w^h in (4.43) that the correction $u^h \leftarrow u^h + u^{2h}$ is still adequate for proper acceleration of Kaczmarz relaxation on level h.

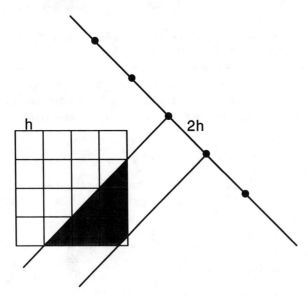

FIG. 4.8. *Image and one projection grid for coarsening of the equations.*

Reducing the cost of coarse-level relaxation sweeps can lead to an improvement in the overall cost of the multigrid process that is proportional to the

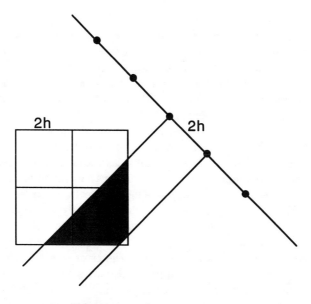

FIG. 4.9. *Fully coarsened image and projection grids.*

number of levels. This usually means a cost savings of a factor proportional to $\log n$. Thus, coarsening the unknowns can be important. But there is potentially much more to gain by coarsening the equations. Assuming that Kaczmarz applied to (4.35) has a worst-case convergence factor $1 - ch^2$ (which is by no means a safe assumption since Kaczmarz's method has not been analyzed in this context, at least to our knowledge), then typical multigrid performance would yield an improvement in speed by a factor proportional to n^2. The main question then is whether the error components that are slow to be reduced by relaxation can be attenuated by coarse-level correction well enough to achieve typical multigrid performance.

To examine this question more closely, consider the related matrix system (4.38) with a given approximation, y, to its solution, y^*. Then the ith step of one Gauss-Seidel sweep applied to y can be written in terms of the algebraic error, $e = y - y^*$, as

$$e \leftarrow e - \frac{\langle w_{(i)}, A\,A^t e \rangle}{\langle w_{(i)}, A\,A^t w_{(i)} \rangle}\, w_{(i)}.$$

Define the *energy error norm* by

$$|||e||| = \langle e, A\,A^t e \rangle^{1/2}.$$

(Actually, this is a norm only if $A\,A^t$ is nonsingular; but we can just consider this argument restricted to the orthogonal complement of the null space of $A\,A^t$, and therefore of A^t, because null space components of A^t have no effect on the true variable $x = A^t y$.) Then the energy error behaves according to

$$|||e|||^2 \leftarrow |||e|||^2 - \alpha_i \langle w_{(i)}, AA^t e \rangle^2,$$

where $\alpha_i = 1/\langle \underline{w}_{(i)}, \underline{A}\underline{A}^t\underline{w}_{(i)}\rangle$. Loosely speaking, if one full Gauss-Seidel sweep is unable to reduce the energy error adequately, then $\underline{A}^t\underline{e}$ must be near (but unequal) to zero in some relative sense. In fact, if α_i is of order one (as it is in our case), then an inadequate convergence rate means that $\|\underline{A}\,\underline{A}^t\underline{e}\|$ must be small relative to $\||\underline{e}\||$. In summary, then, troublesome error components for Kaczmarz's method applied to (4.35) are those that correspond to errors yielding relatively small residuals for the equivalent system, (4.40).

Noting that $L^h(L^h)^*$ maps projections to projections, then one type of error component that yields a relatively small residual can be described physically as a set of projections that backproject to an image that is nearly zero, or *black*. Now, because the projection cell size is no smaller than the image cell size (They are assumed to be equal on level h), no single projection can be backprojected to form a nearly black image unless it is already nearly black itself. However, any two projections can backproject to a near-black image if their gray levels are individually almost constant but of opposite sign. (Note that a negative density makes sense here because we are referring to the error, not the original object.) In fact, two projections of constant opposite densities exactly cancel each other in backprojection, and such pairs generate the null space of $L^h(L^h)^*$. (A block of projections is in the null space of $L^h(L^h)^*$ if each projection is uniformly gray and the sum over the block of their individual total densities is zero.) Thus, the null space of the self-adjoint operator $L^h(L^h)^*$ is in T^{2h}, which is enough to ensure that it cannot impede convergence. But what can be said about the near null space components, which backproject to nearly black images? Here we need only consider components that are orthogonal to the null space of $L^h(L^h)^*$, which means that their individual projections have zero average gray levels. A little reflection reveals that, unless two such interfering projections have nearly the same angle, their individual densities must be smoothly varying. This suggests that near null space components are accurately represented in T^{2h}, which in turn suggests that our coarsening process should be effective.

These heuristics seem to imply that PML based on coarsening only the equations might converge quickly. However, several questions remain. First, two projections at nearly equal angles can have oscillatory individual densities and still interfere to almost cancel in backprojection. Does this mean that PML convergence rates degrade with increasing p and, if so, is there another coarsening process to prevent it? One possible alternative is to redefine T^{2h} as the subspace of T^h consisting of projections that satisfy $f_{\alpha_{2i-1}} = f_{\alpha_{2i}}$, $1 \leq i \leq p/2$. (Here we assume that an appropriate ordering of angles is given, that p is even, and that paired projections have the same number of cells.) A second question is whether all troublesome components can be generated only by interfering pairs of projections, or whether many projections can combine in a substantially different way to nearly cancel on backprojection. Third, as an important practical issue, what is the effect of perturbing f^h out of the range of $(L^h)^*$? We can eliminate potential trouble by forcing each projection to satisfy the compatibility condition, (4.34). While this may not be necessary for Kaczmarz by itself, what is the case when several levels are used? (The form (4.35) should be considered in this context.) Finally, coarsening of the unknown, or image, is compatible with the discretization, but is it really effective or even necessary in practice?

This paradigm shows the clarity that the multilevel projection method provides for its analysis: the singularity of neither L^h nor $(L^h)^*$ matters by itself; the only apparent concern is the "near null space" components of L^h in the range of $(L^h)^*$. On the other hand, conventional methods for image reconstruction that do not take a projection-like approach may be allowing operator singularities to contaminate otherwise well-posed components, which will drastically impair convergence and accuracy. This is especially true for the more traditional methods that model the rays as lines instead of "fat" beams, as we have done. (Of course, fat beam models may not be relevant for many applications.)

We should point out that the roles here identified for equation-coarsening and unknown-coarsening depend on the fact that we are using Kaczmarz as the relaxation process, which means that we are in effect solving (4.40). Another possible relaxation scheme for (4.35) is to apply Gauss-Seidel to the normal equations,

$$(L^h)^* L^h u^h = (L^h)^* f^h,$$

which would have the effect of reversing the roles of the two coarsening processes.

CHAPTER 5

Perspectives

Although modern multigrid research has spanned almost two decades, the field is still wide open. For projection methods in particular, progress is in its early stages. Some of the remaining tasks include developing a theoretical foundation for the basic approach, conducting numerical studies in greater depth to determine its potential, and exploiting more applications to pursue its practical significance. For multilevel methods in general, while the technology has progressed dramatically in the last decade or so for basic applications, the field is yet very much unsettled for the more advanced scientific areas. However, even for basic problems, many challenges remain. For example, we are still in need of: a well-founded and systematic multilevel treatment of advection-dominated systems (e.g., proper discretization and relaxation schemes, particularly for realistic high Reynolds flow models), a sharp and realistic convergence theory for multilevel methods (especially for the V-cycle) that applies to nonlinear and nonvariational systems, and a systematic set of principles for guiding the general design of multilevel processes (e.g., that suggests what subspaces to use).

We do not attempt here in any way to survey open problems in the multilevel discipline; that by itself is a difficult research task. Instead, we focus on just a few of the major questions of special relevance to the material in this monograph. The following list therefore places particular emphasis on projection methods and adaptive refinement. Also, as general forecasts of this kind must be, the brief comments that we make will be rather vague about the specific course that the various research topics might take. Finally, we will not reference other work except to say that most of the relevant results can be found in this monograph, the Frontiers book [McCormick 1989], or the literature that they cite.

a. *High-order FAC*. Develop FAC methods for obtaining high-order accuracy. Basic possibilities include extrapolation-type schemes, higher order approximations (e.g., p-version finite elements or spectral methods), and defect corrections, to name a few. This task could focus on how effective high-order methods are for realistic applications, and how they can be incorporated into fully automatic adaptive schemes.

b. *Self-adaptive FAC*. Study the feasibility of a fully automatic adaptive refinement scheme based on FAC. This should be founded on reliable discretization error estimators, chosen among those used for conventional methods as well as the multilevel estimators enabled by the presence of various levels of approximation. An interesting avenue here is to develop local error estimators for FVE akin to those used with finite element methods. This should be accompanied by theory to validate the applicability of the estimation process. The ultimate goal here should be to develop and study fully automatic multilevel adaptive methods for realistic applications, including time-dependent, three-dimensional fluid flows.

c. *Parallel self-adaptive AFAC*. Using suitable error estimators, develop a self-adaptive refinement scheme that is fully compatible with parallel computation. One of the main concerns here is that the scheme stay within the parallel efficiency range of the AFAC solver: adaptive decision-making, and generation and elimination of grid patches, should not impair the speed of the overall solution procedure. This means that the self-adaptive process should be both local and patch-oriented, with minimal interaction between refinement levels. The ultimate goal here should be to test performance on existing parallel computers.

d. *Time-dependent methods*. MG and FAC methods have been applied to parabolic equations in several different ways. However, the special character of time-dependent equations has not yet been fully exploited, especially in the context of self-adaptive refinement, moving fronts and other phenomena, and full time-parallelization. The objective of this task should be to implement multilevel methods to explore their potential in this regard.

e. *Local grid generation*. Grid generation has generally involved mapping a uniform grid to a global irregular computational domain to simplify treatment of the boundary, for example. One of the problems with this approach is that, to accommodate local irregularities, global complexities are introduced, both in computing the mapping and in transforming the equations. FAC can provide the framework for local grid generation by restricting the mapping process to local patches placed about irregularities, which would limit the added complexity to the regions where it is needed. This task should therefore begin by developing a local grid generation capability for FAC, carefully determining the best way to define and manipulate the resulting composite grid, and testing the approach on irregular boundaries, fronts, shocks, and other phenomena. The objective should be to push the flexibility of FAC to the limit to see how it handles complicated refinement tasks, especially when compared to other approaches like those based on unstructured grids.

f. *Zonal methods*. Most adaptive methods are based on local adjustment of the mesh size or approximation order. However, it may be that numerical solution methods or even PDE models should change from one region to the next. For example, Euler equations can be used in the global domain for a predominantly inviscid flow, while reduced or full Navier-Stokes equations can be used in local zones where viscosity is significant. The multilevel principle that is significant here is that the cruder methods or equations (e.g., Euler equations) should first be solved globally, including the regions where the finer approaches will be taken. This will ensure that the full process converges quickly and that the refinement zone will have good starting guesses. It will also avoid the awkward (and generally deleterious) holes that arise in the global grids when the refinement patches are eliminated. There are many avenues to take here, but an interesting start would be to study advection-diffusion equations, using the diffusion terms in the solvers only in regions where they are relatively significant (e.g., in boundary layers or at turning or stagnation points). Problems of this type can be easily discretized by FVE methods, but deeper questions about the discretization—and the solvers—will likely arise when more sophisticated flow models are considered.

g. *Highly convective flow.* The objective here would be to develop a fully systematic approach to accurate discretization and efficient multilevel solution of flow equations where convection is dominant. This would ultimately involve high Reynolds number flow and other related phenomena. Substantial effort is currently progressing in the development of various difference schemes, finite volume methods, finite element techniques, and other discretization approaches, but no clearly effective and robust choice has yet surfaced. One possible start here is to consider the exponential-type FVE approach, which has preliminarily exhibited acceptable accuracy and achieved fast multigrid performance. Questions that remain here are whether the added cost of exponential-type functions is worth their gain and whether robust methods can be developed on this basic approach. A special challenge would be to construct a theory for these methods that provides realistic discretization error bounds and sharp convergence estimates.

h. *FVE extensions.* The FVE method has been established on a limited theoretical footing. Current theory applies to linear elliptic equations of the form $-\nabla \cdot A\nabla u = f$, where A is a positive-definite symmetric matrix function. A simple convergence theory is obtained by interpreting FVE as an approximate finite element Galerkin scheme. Superconvergence results in a discrete energy norm are obtained by direct analysis of the FVE process. The objective here would be to extend this theory to more sophisticated PDEs, first to advection-diffusion equations, then perhaps to nonlinear elliptic equations, physical conservation laws for systems, and time-dependent PDEs.

i. *Multilevel convergence theory.* Realistic multilevel convergence theory (e.g., that applies to $(1, 1)$ V-cycles and commonly used smoothers) has been obtained only for equations that are essentially linear and self-adjoint. More general cases usually require a sufficient number of relaxation sweeps and apply only to W-cycles and two-grid schemes. A major hurdle in multigrid theory is to extend the mesh-independent $(1, 1)$ V-cycle estimates to a more general setting, including nonlinear and nonself-adjoint PDEs.

j. *PML theory.* Except for the linear and eigenvalue prototypes, no theory for PML exists in any form. One of the first tasks here would be to develop a two-level algebraic convergence theory for PML applied to nonlinear elliptic

PDEs. It would be especially interesting to account for FAS convergence properties and establish under what conditions FAS can provide a suitable approximation to the PML principle. For a general PML convergence theory, a first step would be to obtain local results by linearizing the PDE about the solution, but it would be better to obtain a direct PML result that establishes convergence in a more global sense. Particularly relevant would be a theory that limits the use of asymptotic assumptions (e.g., that the mesh is sufficiently fine), and that obtains analytic FMG results as well as algebraic V-cycle estimates.

k. *Realizability.* The task here is to develop general conditions under which the PML schemes are essentially realizable in both the discretization and coarse-level senses. For cases where they are not, there should be general strategies developed for accurately approximating the discretization and coarsening principles, with theoretical estimates of the errors that this introduces. One of the first steps here is to determine what limits, if any, there are for FAS in this role, or if other strategies suggest themselves. In this context, it would be interesting to compare the performance and robustness of the three basic multilevel strategies for treating nonlinearities: Newton's method with a multigrid scheme applied to the resulting matrix equation, FAS, and PML. Preliminary tests with these three methods on the eigenvalue prototype suggest that they perform very much the same near the solution (which can usually be assured by using nested iteration, or FMG). But the best global convergence properties are exhibited by PML, with FAS often considerably slower and the Newton approach much slower still, even becoming divergent away from the solution. One question is whether this relative performance is indicative of the general case, or just particular to eigenvalue problems.

l. *PML with constraints.* This topic is important, but largely unexplored. Except for the very simple constraint for eigenvalue problems treated in this monograph, and the general constrained optimization problem treated by aggregation in [Gelman and Mandel 1990], multilevel methods of projection type have been considered only for unconstrained variations. The objective here would be to consider the variety of options for treating equality and inequality constraints (e.g., by various projection schemes, penalty functions and Lagrange multipliers, and artificial variables), and to develop both the algorithms and the underlying theory.

m. *Applications.* This is perhaps the widest open research topic. In this monograph we have discussed six paradigms, all in early stages of development. For most of them we have identified several open research questions. But there are many more application areas that have not been very deeply explored by multigrid methods in general, and many more left to be tackled by multilevel projection methods. The list is no doubt very long, but some that come to mind are multiphase flow, Maxwell's equations, magnetohydrodynamics, combustion, and queuing theory and Markov chains.

References

A. Brandt [1984], *Multigrid Techniques: 1984 Guide, with Applications to Fluid Dynamics*, Dept. Appl. Math., Weizmann Inst. of Science, Rehovot 76100, Israel.

W. Briggs [1987], *A Multigrid Tutorial*, Society for Industrial and Applied Mathematics, Philadelphia.

W. Briggs, L. Hart, S. McCormick, and D. Quinlan [1988], *Multigrid methods on a hypercube*, in Multigrid Methods: Theory, Applications, and Supercomputing, S. F. McCormick, ed., Lecture Notes in Pure and Appl. Math., 110, Marcel Dekker, New York, pp. 63–83.

T. Chan and R. Tuminaro [1987], *Design and implementation of parallel multigrid algorithms*, in Multigrid Methods: Theory, Applications, and Supercomputing, S. F. McCormick, ed., Lecture Notes in Pure and Appl. Math., 110, Marcel Dekker, New York, pp. 101–115.

P. Ciarlet [1978], *The Finite Element Method for Elliptic Problems*, North-Holland, Amsterdam.

J. Dongarra, J. Bunch, C. Moler, and G. Stewart [1979], *LINPACK Users' Guide*, Society for Industrial and Applied Mathematics, Philadelphia.

E. Gelman and J. Mandel [1990], *On multilevel iterative methods for optimization problems*, Math. Programming, 48, pp. 1–17.

G. Herman, H. Levkowitz, S. McCormick, and H. Tuy [1982], *Multigrid image reconstruction*, Proc. Workshop on Multiresolution Image Processing and Analysis, Leesburg, VA, A. Rosenfeld, ed., Springer-Verlag, New York.

H. Malcolm Hudson [1990], *Multi-scale reconstruction in image analysis*, Proc. Joint Meeting Bernoulli Soc. for Math. Stat. and Prob. and Inst. for Math. Stat., Uppsala.

K. Ito, S. McCormick, and L. Teijun [1991], *Multilevel Ricatti solvers*, Prelim. Proc. Fifth Copper Mountain Conf. on Multigrid Methods, Copper Mountain, Colorado.

R. Kohn and M. Vogelius [1985], *Determining conductivity of boundary measurements II. Interim results*, Comm. Pure Appl. Math., 38, pp. 644–667.

—— [1987], *Relaxation of a variational method for impedance computed tomography*, Comm. Pure Appl. Math., 40, pp. 745–777.

M. Lemke and D. Quinlan [1991], *Fast adaptive composite grid methods on distributed parallel architectures*, Proc. Fifth Copper Mountain Conf. on Multigrid Methods, Copper Mountain, Colorado.

J. Mandel and S. McCormick [1989], *A multilevel variational method for Au = gBu on composite grids*, J. Comput. Phys., 80, pp. 442–450.

T. Manteuffel, S. McCormick, J. E. Morel, and G. Yang [1990], *Fast multigrid solver for neutron transport problems*, Proc. IMACS First International Conf. on Comp. Physics, Boulder, Colorado.

S. McCormick [1982], *An algebraic interpretation of multigrid methods*, SIAM J. Numer. Anal., 19, pp. 548–560.

—— [1984], *Fast adaptive composite grid (FAC) methods: Theory for the variational case*, in Comp. Suppl. 5, Defect Corr. Methods, K. Böhmer and H. J. Stetter, eds., Springer-Verlag, Vienna.

—— [1985], *A variational theory for multilevel adaptive techniques (MLAT)*, in Multigrid Methods for Integral and Differential Equations, D. J. Paddon and H. Holstein, eds., IMAA Conf. Ser. 3, Clarendon Press, Oxford, pp. 225–230.

——, ed. [1987], *Multigrid Methods*, SIAM Frontiers in Applied Mathematics 3, Society for Industrial and Applied Mathematics, Philadelphia.

—— [1989], *Multilevel Adaptive Methods for Partial Differential Equations*, SIAM Frontiers in Applied Mathematics 6, Society for Industrial and Applied Mathematics, Philadelphia.

S. McCormick and J. W. Ruge [1983], *Unigrid for multigrid simulation*, Math. Comp., 41, pp. 43–62.

S. McCormick and G. Wade [1991], *Multilevel parameter estimation*, Proc. Fifth Copper Mountain Conf. on Multigrid Methods, Copper Mountain, Colorado.

R. V. Southwell [1935], *Stress-calculation in frameworks by the method of "systematic relaxation of constraints,"* parts I, II, Proc. Roy. Soc. (A), 151, pp. 56–91, part III, Proc. Roy. Soc. (A), 153, pp. 41–76.

K. Stüben and U. Trottenberg [1982], *Multigrid methods: Fundamental algorithms, model problem analysis and applications*, in Multigrid Methods, proc. conf. held at Köln-Porz, November 23–27, 1981, W. Hackbusch and U. Trottenberg, eds., Lecture Notes in Mathematics, 960, Springer-Verlag, Berlin, pp. 1–176.

Appendix A: Simple Unigrid Code

Below is a collection of subroutines for performing several cycles of unigrid applied to Poisson's equation with homogeneous Dirichlet boundary conditions.* They apply to the usual 9-point stencil defined on a uniform grid. The coding is simplified by not taking into account the zero structure of the directions used in the iterative process.

```
      subroutine unigrid(n,u,f,r,d,ad,ncyc,nlvl)
c
c
c     This is a simple version of the multigrid simulation scheme, unigrid, which
c     applies to the equation Lu=f, where L=-Laplacian, on the unit square with
c     homogeneous Dirichlet boundary conditions. Here there is no accounting for
c     the zero structure of the directions used in unigrid processing, which makes
c     the code simple but slow to execute.
c
c     This routine performs ncyc (0,1) V cycles of unigrid on an n by n uniform
c     grid using a nested sequence of nlvl grids with mesh sizes differing by a
c     factor of 2. [A (0,1) V cycle uses no relaxation sweeps before coarsening
c     and one after.] n counts interior points only, so the finest mesh size is
c     h=1/(n+1). To ensure proper grid nesting, it is assumed that 2**(nlvl-1)
c     divides n+1. Both arrays u and d are assumed to have zeros as entries that
c     correspond to boundaries (i or j = 0 or n+1). It is assumed that the source
c     term array contains approximations to the continuum function scaled by the
c     factor h*h, that is, the array should approximate the values f(i*h,j*h)*h*h.
c     Note that all arrays are dimensioned with 0:n+1 for convenience. On input,
c     the array u is assumed to contain the initial approximation; on output, it
c     contains the final approximation.
c
c     Subroutine arguments:
c        n       = number of interior x-coordinate [or y-coordinate] grid lines
c        u       = array containing initial/final approximation
c        f       = source term scaled by h*h [h=1/(n+1)]
c        r,d,ad  = work arrays
c        ncyc    = number of V cycles to be performed
c        nlvl    = number of grid levels [2**(nlvl-1) should divide n+1.]
c
      dimension u(0:n+1,0:n+1),f(0:n+1,0:n+1),r(0:n+1,0:n+1)
      dimension d(0:n+1,0:n+1),ad(0:n+1,0:n+1)
c Output the initial Euclidean norm of the residual error.
      call error(n,u,f,r,0)
c Perform ncyc V cycles.
```

* The unigrid routines listed in these appendices can be obtained on request by electronic mail from the author (smccormi@copper.denver.colorado.edu).

103

```
      do 1000 kcyc=1,ncyc
c Cycle from the coarsest grid [lvl=1] to the finest [lvl=nlvl].
      do 100 lvl=1,nlvl
c Define the mesh factor m and the number ncoarse of x-coordinate lines for
c level lvl.
      m=2**(nlvl-lvl)
      ncoarse=(n+1-m)/m
c Begin one sweep on level lvl using the hat function centered at point (i,j).
c [The corresponding coarse grid point is (idel,jdel).]
      do 80 idel=1,ncoarse
      do 80 jdel=1,ncoarse

      i=m*idel
      j=m*jdel
c Compute the residual r=Au-f, where A is the matrix corresponding to the nine
c point stencil discretization of the differential equation Lu=f.
      call amult(n,u,r)
      call xpcy(n,r,f,-1.)
c Compute the direction d, the matrix-vector product Ad, and the inner products
c <d,r> and <d,Ad>.
      call direction(n,d,i,j,m)
      call amult(n,d,ad)
      call dot(n,d,r,ddotr)
      call dot(n,d,ad,ddotad)
c Compute the step size s and update the vector u by the correction s*d.
      s=-ddotr/ddotad
      call xpcy(n,u,d,s)
   80    continue
  100    continue
c Output the Euclidean norm of the residual error for cycle number kcyc.
      call error(n,u,f,r,kcyc)
 1000 continue
      return
      end

      subroutine amult(n,u,au)
c
c
c Compute the matrix-vector product Au and store in array au.
c
      dimension u(0:n+1,0:n+1),au(0:n+1,0:n+1)
      do 20 i=1,n
      do 20 j=1,n
      au(i,j)=9.*u(i,j)
      do 10 k=-1,1
      do 10 l=-1,1
      au(i,j)=au(i,j)-u(i+k,j+l)
   10    continue
      au(i,j)=au(i,j)/3.
   20 continue
      return
      end

      subroutine direction(n,d,i,j,m)
c
c Compute the hat function centered at point (i,j) with support at points
c (i+k,j+l), where k,l=1-m,2-m,...,-1,0,1,...,m-2,m-1.
c
      dimension d(0:n+1,0:n+1)
      call zero(n,d)
      do 10 k=1-m,m-1
      do 10 l=1-m,m-1
      d(i+k,j+l)=(m-abs(k))*(m-abs(l))
   10 continue
      return
      end

      subroutine dot(n,x,y,xdoty)
c
c
c Compute the dot product of arrays x and y and store in xdoty.
c
      dimension x(0:n+1,0:n+1),y(0:n+1,0:n+1)
      xdoty=0.
      do 10 i=1,n
      do 10 j=1,n
      xdoty=xdoty+x(i,j)*y(i,j)
   10 continue
```

```
      return
      end

      subroutine error(n,u,f,r,kcyc)
c
c
c Output the Euclidean norm of the residual r=Au-f for cycle number kcyc.
c
      call amult(n,u,r)
      call xpcy(n,r,f,-1.)
      call dot(n,r,r,err)
      err=sqrt(err)
      write(*,*)'residual error for cycle ',kcyc,' = ', err
      return
      end

      subroutine xpcy(n,x,y,c)
c
c
c Form the linear combination x+c*y for the vectors x,y and scalar c and
c store in x.
c
      dimension x(0:n+1,0:n+1),y(0:n+1,0:n+1)
      do 10 i=1,n
      do 10 j=1,n
        x(i,j)=x(i,j)+c*y(i,j)
10 continue
      return
      end

      subroutine zero(n,x)
c
c
c Zero out the vector x.
c
      dimension x(0:n+1,0:n+1)
      do 10 i=0,n+1
      do 10 j=0,n+1
        x(i,j)=0.
10 continue
      return
      end
```

Appendix B: More Efficient Unigrid Code

Below is a collection of subroutines for performing several cycles of unigrid applied to Poisson's equation with homogeneous Dirichlet boundary conditions. They apply to the usual 9-point stencil defined on a uniform grid. The coding is complicated by its accounting for the zero structure of the directions used in the iterative process, but this greatly improves efficiency.

```
      subroutine unigrid(n,u,f,d,ncyc,nlvl)
c
c
c     This is a relatively efficient version of the multigrid simulation scheme,
c     unigrid, which applies to the equation Lu=f, where L=-Laplacian, on the unit
c     square with homogeneous Dirichlet boundary conditions. Here account is made
c     of the zero structure of the directions used in unigrid processing, which
c     makes the code a little complicated but greatly improves execution speed.
c     This is done by generating a reference hat function in the lower left-hand
c     corner of the region, then generating hat functions centered at other grid
c     points by a translation of indices.
c
c     This routine performs ncyc (0,1) V cycles of unigrid on an n by n uniform
c     grid using a nested sequence of nlvl grids with mesh sizes differing by a
c     factor of 2. [A (0,1) V cycle uses no relaxation sweeps before coarsening
c     and one after.] n counts interior points only, so the finest mesh size is
c     h=1/(n+1). To ensure proper grid nesting, it is assumed that 2**(nlvl-1)
c     divides n+1. Both arrays u and d are assumed to have zeros as entries that
c     correspond to boundaries (i or j = 0 or n+1). It is assumed that the source
c     term array contains approximations to the continuum function scaled by the
c     factor h*h, that is, the array should approximate the values f(i*h,j*h)*h*h.
c     Note that all arrays are dimensioned with 0:n+1 for convenience. On input,
c     the array u is assumed to contain the initial approximation; on output, it
c     contains the final approximation.
c
c     Subroutine arguments:
c        n     =  number of interior x-coordinate [or y-coordinate] grid lines
c        u     =  array containing initial/final approximation
c        f     =  source term scaled by h*h [h=1/(n+1)]
c        d     =  work array
c        ncyc  =  number of V cycles to be performed
c        nlvl  =  number of grid levels [2**(nlvl-1) should divide n+1.]
c
      dimension u(0:n+1,0:n+1),f(0:n+1,0:n+1),d(0:n+1,0:n+1)
c     Output the initial Euclidean norm of the residual error.
      call error(n,u,f,0)
```

107

```
c  Perform ncyc V cycles.
         do 1000 kcyc=1,ncyc
c  Cycle from the coarsest grid [lvl=1] to the finest [lvl=nlvl].
         do 100 lvl=1,nlvl
c  Define the mesh factor m and the number ncoarse of x-coordinate lines for
c  level lvl.
         m=2**(nlvl-lvl)
         ncoarse=(n+1-m)/m
c  Define the template direction vector d: the hat function for level lvl that
c  is centered at a grid point (i,j) that corresponds to a coarse grid point

c  has entries in position (i+k,j+l) equal to d(m+k,m+l), where k,l=1-m,2-,...,
c  -1,0,1,...,m-2,m-1, and equal to zero elsewhere.
         call direction(n,d,m)
c  Begin one sweep on level lvl using the hat function centered at point (i,j).
c  [This corresponds to coarse grid point (idel,jdel).]
         do 80 idel=1,ncoarse
         do 80 jdel=1,ncoarse
         i=m*idel
         j=m*jdel
c  Compute the residual r=Au-f, where A is the matrix corresponding to the nine
c  point stencil discretization of the differential equation Lu=f. Compute the
c  matrix-vector product Ad and the inner products <d,r> and <d,Ad>.
         ddotr=0.
         ddotad=0.
         do 40 k=1-m,m-1
         do 40 l=1-m,m-1
           r=9.*u(i+k,j+1)-f(i+k,j+1)
           ad=9.*d(m+k,m+1)
           do 20 kdel=-1,1
           do 20 ldel=-1,1
             r=r-u(i+k+kdel,j+1+ldel)
             ad=ad-d(m+k+kdel,m+1+ldel)
   20        continue
           r=r/3.
           ad=ad/3.
           ddotr=ddotr+d(m+k,m+1)*r
           ddotad=ddotad+d(m+k,m+1)*ad
   40    continue
c  Compute the step size s and update the vector u by the correction s*d.
         s=-ddotr/ddotad
         do 60 k=1-m,m-1
         do 60 l=1-m,m-1
           u(i+k,j+1)=u(i+k,j+1)+s*d(m+k,m+1)
   60    continue
   80    continue
  100    continue
c  Output the Euclidean norm of the residual error for cycle number kcyc.
         call error(n,u,f,kcyc)
 1000 continue
      return
      end

      subroutine direction(n,d,m)
c
c
c  Compute the template direction d as the reference hat function defined on
c  the n by n fine grid, centered at point (m,m), and nonzero at points (i,j),
c  where i,j=1,2,...,2*m-1.
c
         dimension d(0:n+1,0:n+1)
         do 10 i=0,2*m
         do 10 j=0,2*m
           d(i,j)=(m-abs(m-i))*(m-abs(m-j))
   10    continue
      return
      end

      subroutine error(n,u,f,kcyc)
c
c
c  Output the Euclidean norm of the residual r=Au-f for cycle number kcyc.
c
         dimension u(0:n+1,0:n+1),f(0:n+1,0:n+1)
         err=0.
         do 20 i=1,n
         do 20 j=1,n
```

```
        r=9.*u(i,j)-f(i,j)
        do 10 k=-1,1
        do 10 l=-1,1
          r=r-u(i+k,j+l)
   10   continue
        r=r/3.
        err=err+r*r
   20 continue
      err=sqrt(err)
      write(*,*)'residual error for cycle ',kcyc,' = ', err
      return
      end

      subroutine zero(n,x)
c
c
c Zero out the vector x.
c
      dimension x(0:n+1,0:n+1)
      do 10 i=0,n+1
      do 10 j=0,n+1
        x(i,j)=0.
   10 continue
      return
      end
```

Appendix C: Modification to Unigrid Code for Local Refinement

Below is a listing of the code segment for modifying the global-grid unigrid routine given in Appendix A. It is to be used to convert the finest level to a local grid patch covering the lower left-hand quadrant of the unit square.

Immediately after the dimension statements in the unigrid subroutine, include:

```
            call zero(n,u)
            do 2 i=1,(n-3)/4
            do 2 j=1,(n-3)/4
              call direction(n,d,i,j,1)
              call xpcy(n,u,d,ran(1))
2           continue
            do 4 i=2,n-1,2
            do 4 j=2,n-1,2
              call direction(n,d,i,j,2)
              call xpcy(n,u,d,ran(1))
4           continue
            call amult(n,u,f)
            call zero(n,u)
```

Immediately after the statement defining the variable ncoarse in the unigrid subroutine, include:

```
            if(lvl.eq.nlvl)ncoarse=(n-3)/4
```

[This assumes that 4 divides n-3.]

Index